Inventing the cave man

Manchester University Press

STUDIES IN POPULAR CULTURE

General editor: Professor Jeffrey Richards
Already published

Dancing in the English style: consumption, Americanisation, and national identity in Britain, 1918–50
Allison Abra

Christmas in nineteenth-century England
Neil Armstrong

Healthy living in the Alps: the origins of winter tourism in Switzerland, 1860–1914
Susan Barton

Working-class organisations and popular tourism, 1840–1970
Susan Barton

Leisure, citizenship and working-class men in Britain, 1850–1945
Brad Beaven

Leisure and cultural conflict in twentieth-century Britain
Brett Bebber (ed.)

Leisure cultures in urban Europe, c.1700–1870: a transnational perspective
Peter Borsay and Jan Hein Furnée (eds)

British railway enthusiasm
Ian Carter

Railways and culture in Britain
Ian Carter

Time, work and leisure: life changes in England since 1700
Hugh Cunningham

Darts in England, 1900–39: a social history
Patrick Chaplin

Holiday camps in twentieth-century Britain: packaging pleasure
Sandra Trudgen Dawson

History on British television: constructing nation, nationality and collective memory
Robert Dillon

The food companions: cinema and consumption in wartime Britain, 1939–45
Richard Farmer

Songs of protest, songs of love: popular ballads in eighteenth-century Britain
Robin Ganev

Heroes and happy endings: class, gender, and nation in popular film and fiction in interwar Britain
Christine Grandy

Women drinking out in Britain since the early twentieth century
David W. Gutzke

The BBC and national identity in Britain, 1922–53
Thomas Hajkowski

From silent screen to multi-screen: a history of cinema exhibition in Britain since 1896
Stuart Hanson

Juke box Britain: Americanisation and youth culture, 1945–60
Adrian Horn

Popular culture in London, c. 1890–1918: the transformation of entertainment
Andrew Horrall

Popular culture and working-class taste in Britain, 1930–39: a round of cheap diversions?
Robert James

The experience of suburban modernity: how private transport changed interwar London
John M. Law

Amateur film: meaning and practice, 1927–1977
Heather Norris Nicholson

Films and British national identity: from Dickens to *Dad's Army*
Jeffrey Richards

Cinema and radio in Britain and America, 1920–60
Jeffrey Richards

Looking North: Northern England and the national imagination
Dave Russell

The British seaside holiday: holidays and resorts in the twentieth century
John K. Walton

Politics, performance and popular culture in the nineteenth century
Peter Yeandle, Katherine Newe and Jeffrey Richards

Inventing the cave man

From Darwin to the Flintstones

ANDREW HORRALL

Manchester University Press

Copyright © Andrew Horrall 2017

The right of Andrew Horrall to be identified as the author of this work has been asserted by him in accordance with the Copyright, Designs and Patents Act 1988.

Published by Manchester University Press
Altrincham Street, Manchester M1 7JA

www.manchesteruniversitypress.co.uk

British Library Cataloguing-in-Publication Data
A catalogue record for this book is available from the British Library

ISBN 978 1 5261 1384 9 hardback

First published 2017

The publisher has no responsibility for the persistence or accuracy of URLs for any external or third-party internet websites referred to in this book, and does not guarantee that any content on such websites is, or will remain, accurate or appropriate.

Typeset in 10.5/14pt Garamond by
Servis Filmsetting Ltd, Stockport, Cheshire
Printed in Great Britain by
TJ International Ltd, Padstow

STUDIES IN POPULAR CULTURE

There has in recent years been an explosion of interest in culture and cultural studies. The impetus has come from two directions and out of two different traditions. On the one hand, cultural history has grown out of social history to become a distinct and identifiable school of historical investigation. On the other hand, cultural studies has grown out of English literature and has concerned itself to a large extent with contemporary issues. Nevertheless, there is a shared project, its aim, to elucidate the meanings and values implicit and explicit in the art, literature, learning, institutions and everyday behaviour within a given society. Both the cultural historian and the cultural studies scholar seek to explore the ways in which a culture is imagined, represented and received, how it interacts with social processes, how it contributes to individual and collective identities and world views, to stability and change, to social, political and economic activities and programmes. This series aims to provide an arena for the cross-fertilisation of the discipline, so that the work of the cultural historian can take advantage of the most useful and illuminating of the theoretical developments and the cultural studies scholars can extend the purely historical underpinnings of their investigations. The ultimate objective of the series is to provide a range of books which will explain in a readable and accessible way where we are now socially and culturally and how we got to where we are. This should enable people to be better informed, promote an interdisciplinary approach to cultural issues and encourage deeper thought about the issues, attitudes and institutions of popular culture.

Jeffrey Richards

For Violet

Contents

List of illustrations	*page* x
General editor's foreword	xiii
Acknowledgements	xv
Introduction	1
1 Mass culture: the Victorian world picture	10
2 Darwin, Du Chaillu and Mr Gorilla: the lions of the season	30
3 The parents of Adam and Eve: missing links	57
4 Antediluvian pictorial fun: E.T. Reed and the prehistoric peeps	80
5 He of the auburn locks: George Robey, the Edwardian cave man	115
6 Cave dwellers of Flanders: the First World War	147
7 Modern times: the Victorian cave man's long afterlife	170
Conclusion	198
Bibliography	203
Index	215

Illustrations

1 George Robey as the prehistoric man, c. 1902. © National Portrait Gallery, London. *page 3*
2 Gorilla acrobats, *Fun* (10 March 1886). Reproduced by kind permission of the Syndics of Cambridge University Library. 42
3 'Gorilla Britannicus', *Vanity Fair* (8 February 1862). Reproduced by kind permission of the Syndics of Cambridge University Library. 45
4 An irreverent take on Farini and Krao, *Illustrated Sporting and Dramatic News* (6 January 1883). © The British Library Board (General Reference Collection HIU.LD53). 70
5 The discovery of the missing link, *Moonshine* (19 January 1895). © The British Library Board (General Reference Collection P.P. 5272.p). 82
6 Charles Shannon, prehistoric Wimbledon, *Universal Review* (March 1889). Reproduced by kind permission of the Syndics of Cambridge University Library. 86
7 E.T. Reed, a night lecture on evolution, *Punch* (23 June 1894). Reproduced by kind permission of the Syndics of Cambridge University Library. 93
8 E.T. Reed, prehistoric Fashoda, *Punch* (15 October 1898). Reproduced by kind permission of the Syndics of Cambridge University Library. 94
9 E.T. Reed, prehistoric Henley Regatta (1894), as reprinted in E.T. Reed, *Prehistoric Peeps* (London: Bradbury and Agnew, 1896). 99
10 Prehistoric polo, *Graphic* (7 September 1895). Reproduced

	by kind permission of the Syndics of Cambridge University Library.	102
11	A racist view of prehistory, *Dundee Evening Post* (21 August 1901). © The British Library Board (General Reference Collection MFM.M58153–6).	103
12	Prehistoric sports in Australia, *Sydney Mail* (28 March 1906). National Library of Australia, Bib ID 2318597.	120
13	A prehistoric political cartoon, *Hull Daily Mail* (24 April 1905). © The British Library Board (General Reference Collection, MFM.M88252–54).	122
14	A prehistoric Lord Mayor's show, *Illustrated London News* (9 March 1907). Reproduced by kind permission of the Syndics of Cambridge University Library.	125
15	Prehistoric umbrella, *Burnley Gazette* (25 September 1907). © The British Library Board (Newspapers No. 58–4170).	126
16	Prehistoric motoring, Avon Tyres, *Tatler* (20 March 1912). Reproduced by kind permission of the Syndics of Cambridge University Library.	128
17	A prehistoric party-goer: the Jewish feast of Purim in Finsbury, *Tatler* (29 March 1905). Reproduced by kind permission of the Syndics of Cambridge University Library.	130
18	Prehistory in film: Seymour Hicks as King Mugslot and Jessie Fraser as Coral, *Sketch* (29 July 1914). Reproduced by kind permission of the Syndics of Cambridge University Library.	149
19	Prehistoric military medicine, *Searchlight* (October 1916). Reproduced by kind permission of the Syndics of Cambridge University Library.	157
20	Prehistoric battlefield tactics, *Gnome* (August 1917). Reproduced by kind permission of the Syndics of Cambridge University Library.	158
21	An early depiction of cave women, *Thistle: Scottish Women's Hospitals for Foreign Service Souvenir Book* (Glasgow: John Horn, 1916). Reproduced by kind permission of the Syndics of Cambridge University Library.	159
22	H.G. Wells and Ray Lankester menaced by a dinosaur, *Bystander* (24 December 1919). Reproduced by kind permission of the Syndics of Cambridge University Library.	171
23	A prehistoric Scout leader, *Sketch* (9 July 1913). Reproduced	

	by kind permission of the Syndics of Cambridge University Library.	173
24	Prehistoric chewing gum, *Age* (10 October 1927). National Library of Australia, Bib ID 2908913.	175

General editor's foreword

The popular images of many groups from the distant past – the Druids and the Vikings for instance – still exist today in the forms essentially invented in the eighteenth and nineteenth centuries. Cavemen are another such group and they are the subject of Andrew Horrall's fascinating, wide-ranging and immensely enjoyable new book. The discovery in the 1850s and 1860s of Neanderthal and Cro-Magnon skeletons, collectively dubbed 'cave men' by Sir John Lubbock in 1865, created an appetite for studies, stories and speculation. This coincided with the evolution debate sparked by the publication in 1859 of Charles Darwin's *On the Origin of Species*. The mass entertainment industry that emerged in the nineteenth century fed on both currents of interest. Horrall examines in detail the most significant manifestations of these interests and their popular influence not just in Britain but also in America and the British Empire.

He begins with the search for the 'missing link', the putative intermediate stage between apes and humans. There was the 'gorilla craze' of the 1860s when Paul Du Chaillu's displays of stuffed gorillas fuelled speculation about the link. Later a live gorilla, Mr Pongo, was displayed by flamboyant showman Guillermo Farini and proclaimed the 'missing link' in the 1870s. After Pongo died, Farini turned up another link, Krao, a hairy female discovered in Burma. The 'missing link' entered the popular vocabulary, as evidenced by advertisements for Monkey Brand soap ('the missing link in household cleanliness'). E.T. Reed's 'prehistoric peeps' cartoons in *Punch* in the 1890s and 1900s, used the archetypal cave man, with his unruly hair, animal hide clothing and stone axe, to satirise contemporary institutions, politicians and leisure activities. His approach was much imitated, not least by the American Frederick Burr Opper's cartoon series 'Our antediluvian ancestors' in the *New York Evening Journal*.

The comedian George Robey's music hall sketch 'The prehistoric man' involved a romantic triangle in which two cave men competed for the affections of a cave woman. Robey and Reed established an image which was endlessly repeated in books, magazines, stage shows, gramophone records, cartoons, pageants and later films. The First World War introduced a new take on the iconography with propaganda claiming that the Germans had reverted to prehistoric savagery but the British and their allies embodied the fighting spirit of the cave men. From as early as 1905 films mined the prehistoric theme with contributions from such screen notables as D.W. Griffith (*Man's Genesis*, 1912), Charlie Chaplin (*His Prehistoric Past*, 1914) and Buster Keaton (*Three Ages*, 1923). All three featured the romantic triangle theme of Robey's sketch. Horrall concludes with an account of the long afterlife of the Victorian cave man, as he takes us through a remarkable range of films from *One Million BC* (1940) and *Teenage Caveman* (1958) to *The Flintstones* (1960–66) and *The Clan of the Cave Bear* (1986). Throughout he demonstrates that the cave man imagery reinforced contemporary views of gender, race, class and nationality. The whole adds up to a triumph of original research, interpretative analysis and contextual social history.

Jeffrey Richards

Acknowledgements

I researched much of this book online, accessing abundant sources with equal ease while living in The Hague and in Ottawa. This was a novelty, because I am personally and professionally attracted to dust-begrimed documents. But my greatest debts are now to the organisations and institutions that have digitised so many historic newspapers, periodicals, archival records, images and journals. The potential to access and understand the past has been revolutionised.

Not everything is digital. I am fortunate to work at Library and Archives Canada, which holds the country's most extensive research collections, and to have access to the British Library's unparalleled resources. Poring over original documents among fellow researchers in the reading rooms of such institutions is stimulating in ways – romantic though they may be – that must be protected against a headlong belief that everything important has been or will ever be digitised. I have also been greatly helped by the interlibrary loans staff of the Ottawa Public Library, who located ever more obscure books about cave men, and more recently by Domniki Papadimitriou of the Cambridge University Library, who was endlessly helpful with images.

I would recommend Manchester University Press to anyone.

If I carried out much of my work through the internet's proxy, two people have always been physically present. I am grateful to them above all:

Amy. My friend, companion and love.
Violet. You make me prouder and happier every day.

The book is done. As a cave man once said: 'Yabba-dabba-doo!'

Introduction

Captain Robert Falcon Scott and the members of his second expedition to the South Pole established a base camp on the Antarctic coast in January 1911. Twenty-five men lived for the next year in a prefabricated building whose interior walls were made from crates bearing the names of iconic Edwardian manufacturers – Fry's Cocoa, Huntley and Palmer's Biscuits, Lyle's Golden Syrup, Tate's Sugar, Sunlight Soap, Bovril, Oxo and Heinz. After caching stores and supplies along the route that would be used for the springtime trek to the Pole, the men waited out the winter darkness. Chores consumed most of their time, though they also held football matches in the snow, attended lectures and lantern slide shows, read through an extensive library, played tunes on a pianola and produced a comic newspaper. Scott recorded in his diary that amid this comradely activity, after dinner each night 'the gramophone is usually started by some kindly disposed person'.[1] The Gramophone Company had donated the machine along with hundreds of recordings, ranging from the patriotic strains of the National Anthem and military airs to the Norse heroism evoked by Richard Wagner's 'Ride of the Valkyries' and the levity of operettas and music hall favourites.

As another member of the expedition noted, out of all these choices, '[George] Robey on "Golf" and the "Prehistoric man" are very popular.'[2] The latter was a short sketch about a cave man's amorous misadventures that Robey, one of the era's most popular comedians, had toured since 1902. We can imagine Robey's voice emerging through the ethereal cackle of early recordings to evoke comforting memories of home as much as the labels on the boxes that surrounded Scott's men. Robey would also have conjured up the comic cave men that had been commonly seen in cartoons, stories, songs, plays, pageants, parades and poems since the mid-1890s. These scruffy prehistoric humans inhabited a fanciful historical epoch, depicted as an archaic version of contemporary Britain. The explorers, who hailed from the centre of the British Empire as well as its Australian, Canadian and Caribbean peripheries, would have known about this vision of Britain's most distant past, thanks to the way in which it had been exported throughout the globe. The men may have joked about how their sun- and wind-marked faces, ragged hair and heavy fur, leather and wool garments made them look prehistoric. They may also have smiled that they had created an archaic approximation of the Royal Navy, one of Britain's most venerable and venerated institutions, in a timber hut at the bottom of the earth: walls of crates demarcated the officer's quarters from those of the men, and Scott dined at the head of the table, though everyone used enamelled metal plates and mugs and sat beneath the bulky equipment that hung from the rafters. The gramophone, whose gleaming horn of inlaid wood exemplified middle-class domesticity, must have looked conspicuously out of place.

In early 1912, a party led by Scott reached the Pole, only to die on the homeward journey. Scott's diary, which rang with fortitude, duty and courage, enshrined him as one of the Edwardian era's greatest heroes and inured him to post-imperial reappraisals. Robey on the other hand never successfully adapted to film as it emerged in the 1920s, and so he has been consigned to the world of specialist collectors. Though Robey's 'Prehistoric man' sketch was one of the most popular and influential music hall turns ever devised, his biographers have paid it scant attention.[3] Such is the oblivion into which it has fallen that photographs of Robey in costume are believed to show him in the role of Robinson Crusoe (see Figure 1).[4]

Robey's obscurity reflects a wider ignorance about the history of prehistoric characters in mass culture, even though modern audiences are just as enrapt with clumsy and unintelligent cave men as Scott's explorers had been. These prehistoric humans comically confront archaic versions of the contemporary world as a way of satirising its institutions, ideals and beliefs. It is a fundamentally unscientific vision in which dinosaurs and humans coexist, though they were actually separated by aeons. Yet it is an extremely influential view of the distant past. Satellite communications make it hard to imagine a twenty-first-century group being as isolated as Scott's expedition. But a modern expedition is just as likely to pack music, films and television programmes. Such a collection might well include *The Croods*, the hugely successful 2013 Hollywood prehistoric cartoon. Much

George Robey as the prehistoric man, c. 1902.

as Robey's sketch had, the film would provide the isolated party with a comforting reminder of urban life. Their response to *The Croods* would be conditioned, just as that of Scott's men had been, by long exposure to prehistoric comedies featuring archaic technology like foot-powered cars and domesticated dinosaurs. Robey epitomised this world to the Edwardians, while modern explorers might point to *The Flintstones*, one of the most successful animated television series of all time.

Like virtually all twenty-first-century viewers, the explorers would be unaware that the show's protagonist, Fred Flintstone, descends directly from Robey's cave man. The comic view of the deep past that inspired both characters is an enduring conceit in global mass culture and has been more influential in shaping popular ideas about prehistory than scientific theories, cave art and the fossil remains of ancient hominids. Constance Areson Clark has argued that 'people learn a lot of things about evolution from cartoons, and they know some of these things without thinking much about them – because they learned them from cartoons'.[5] A fair amount has been written about pictorial and literary attempts to scientifically recreate prehistory, and a great deal about dinosaurs, but only the most basic history of cave man characters has been sketched.[6] This oversight is odd, because from the time when proto-human fossils were first discovered in the mid-nineteenth century, writers, illustrators and performers imaginatively recreated the world these creatures had inhabited. Prehistoric humans were initially depicted with explicitly simian features, encapsulating the evolutionary threat that consanguinity with apes posed to Victorian social, religious and racial ideals. Popular visual and stage representations of prehistoric creatures were more simple than scientific or literary ones, but they were not simplistic. They incorporated commonly understood, though often bowdlerised, ideas about human prehistory that had been spread through politically radical magazines, lectures at mechanics' institutes, fairgrounds and popular theatres. As Bernard Lightman argued, such popular activity 'can actively produce its own indigenous science, or can transform the products of elite culture'.[7] In this case, by the end of the century, recognisably modern cave men had emerged, while the scientific context in which they were first depicted had been leavened with historical knowledge of Britain's most distant past.

Robey's cave man sketch was the apotheosis of these broad, popular evocations of evolutionary themes – from gorillas and simian missing links to quasi-human cave men. The frequency with which they were incorporated in print, song, images, theatre and music hall had been amplified by the publication of Charles Darwin's evolutionary treatise *On the Origin of Species* in 1859. Few books have been so heavily scrutinised. The centenary of *Origin*'s publication prompted scholars to assess Darwin's place in evolutionary debates, spawning an academic industry that continues unabated more than fifty years later. A broad consensus now contextualises *Origin* within a long series of evolutionary revelations, attributing its pre-eminence to the efforts of well-placed champions and to the ways in which science was popularised for a mass public. Without attempting to reassert Darwin's ascendancy, this book builds on Alvar Ellegård's early and important investigation of 'how information and opinions concerning [*Origin*] spread throughout the social fabric of mid-Victorian Britain'.[8] It does so by charting a stream flowing through the heart of popular culture, whose headwaters – an appropriate metaphor for a story that begins during the great age of exploration – are located in the clamorous debates that greeted the book.

As we shall see, prehistoric characters evolved – readers should brace for the odd, almost unavoidable evolutionary pun in this book – from aggressive beasts, whose existence threatened mid-Victorian social, scientific and religious beliefs, into recognisably modern humans depicted in situations that poked fun at the contemporary world. Interest was initially generated by showmen attempting to profit from scientific discoveries, especially when professionals argued in public over the importance of a new find or theory. Changes in the way in which prehistoric characters were portrayed reflected the pressures on amateur and professional showmen, artists and illustrators to constantly update their works. Public fascination peaked with each new find, while the subsequent troughs were filled by stubborn performers, writers and illustrators who eked diminished livings from their unchanging prehistoric allusions. As a result, the public encountered popular manifestations of prehistory throughout the rest of the century, showing how deeply evolutionary ideas had penetrated the mass imagination.

One of the most evocative concepts in human prehistory is that of the 'missing link', the extremely ancient species at the junction where ape and human lines diverged. Scientists have long since dropped the idea that the missing link was a single creature, as the branches on the evolutionary tree have become ever more numerous and dense. It is similarly reductive to identify an exact moment at which the modern cave man was created, because mass culture casually and continuously melded high and low ideas in a complex tangle of printed texts, images, music and theatre. However, the role of missing link is played in this book by the immensely popular series of cartoons entitled 'Prehistoric peeps' that first appeared in the comic magazine *Punch* in December 1893. They were drawn by Edward Tennyson Reed, who is memorialised in *The Dictionary of National*

Biography rather preciously as 'probably the originator of antediluvian pictorial fun'.[9] Reed was not the first artist to depict the prehistoric world, but his drawings marked the point at which its inhabitants forever ceased being the threatening, simian products of an evolutionary nightmare and became the creatures we know today: comic, club-wielding modern humans with dishevelled hair and rough hide clothing, whose world is an archaic version of our own. Reed's vision of prehistoric Britain was soon being recreated in rival publications, in fiction, on stage and in the historical pageants and parades that proliferated at the turn of the twentieth century. These accorded the 'peeps', as Reed's prehistoric humans were known, a status akin to Britannia and John Bull as imaginative evocations of the British national character, and the epoch in which they lived became a popularly recognised episode in British history. Reed's aesthetic and sense of humour also migrated from London across the globe, either interbreeding with or displacing localised evocations of the ancient past.

At the turn of the twentieth century, Reed's idea was also taken up by American cartoonists, who recalibrated its comic sensibilities for their audiences. They did so as the United States emerged as the cultural centre of the English-speaking world, thereby expanding and consolidating cave men in the global consciousness. The cave man took so long to establish itself in America for two reasons. Firstly, Americans had been enthralled by the giant, ancient extinct mammoths and dinosaurs that had first been discovered and displayed there in the colonial period. The animals had been coupled to a nationalist political narrative as evidence of the continent's physical grandeur and exceptionalism. More importantly, *Origin* was published in New York amid increasingly angry and violent disputes about the ongoing existence of slavery; a context that made suggestions that humans had descended from African apes explosive. Inhabitants of the industrialising north-east felt slavery was a repugnant anachronism, while southerners argued that their agrarian economy would founder without it. Political, physical and moral battles culminated in the 1860 presidential election, in which Abraham Lincoln, whose Republican Party had been founded less than a decade before on an anti-slavery platform, was elected despite virtually no southern support. This precipitated the start of the Civil War in April 1861. Peace brought decades of national reconstruction in which academics and theologians struggled to find an accommodation between religion and science. Nonetheless, by about 1900, American illustrators, inspired by but not slavishly imitating British models, had begun drawing the innocuous comic cave man character that is familiar to contemporary audiences.

I refer to the denizens of this comic world as 'cave men' for two reasons. Firstly, the term has no precise scientific meaning, despite having been coined by the archaeologist Sir John Lubbock in 1865 to identify the ancient humans whose remains, tools and art had been found across Europe intermixed with the bones of extinct mammals.[10] In this book, the term denotes a creature that fancifully incorporates palaeontological evidence, Classical depictions of hide-clad, club-swinging wild men and broader Western artistic visions of the ancient past. As such, these cave men reside in the popular imagination, rather than in scientific texts or museum exhibits. They are also overwhelmingly male, inhabiting a world that is virtually always a conservative projection of contemporary society in which gender roles are unchallenged. Cave women are most commonly secondary characters, depicted as shrewish wives or eye-batting and increasingly hyper-sexualised objects of lust. While I am conscious that employing an explicitly masculine term risks replicating twenty-first-century linguistic gender norms, it seems to my ear to be the most evocative way of identifying the characters – male and female, old and young – that are the subjects of this study.

This book draws together references to prehistoric creatures – ranging from the scientifically sound to the satirical – from a broad range of newspapers, magazines and other publications from throughout Britain, the empire and the United States. Allusions appeared frequently and prominently in some publications and almost never, or only as tiny asides, in others. These are traceable because huge numbers of historic newspapers, magazines and books have been digitised and are full-text searchable. The superabundant sources are overwhelming, making it impossible to claim that the widespread evocations of prehistory presented in this book are definitive or exhaustive. Such a job is best left to particularly determined indexers and bibliographers.[11] However, the sources show that interest in the prehistoric ancestors of modern humans was expressed through an increasingly standardised set of references and visual iconography. The absolute wealth of material has also made it difficult to select images. In every instance, I have endeavoured to include ones that illustrate particularly important points or that represent broader ideas or artistic genres. Readers wishing to see any other images discussed in this book should be able to find them in the same ever-expanding databases.

Like every missing link unearthed on Java, Ethiopia, the Olduvai Gorge or the Sterkfontein, my evidence may eventually be displaced by researchers fossicking through newly digitised collections. They may discover a slightly older and more compelling comic vision of prehistory in Jarrow, Etchingham, Oldham, Stepney or some other understudied locality. Nonetheless, I am confident that any new evidence will demonstrate that the modern global cave man descends from simian forebears who emerged in the mid-Victorian popular imagination and roamed, at least metaphorically, throughout Britain and beyond, displacing earlier visions as they did. So I present this book in the spirit of a palaeontologist, aware that further discoveries may shift and complexify the origins and evolution of the modern cave man, while leaving the overall path of its descent intact.

Notes

1 Captain Robert Falcon Scott's journal for 19 June 1911, as published in Peter King (ed.), *Scott's Last Journey* (London: Harper Collins, 1999), p. 103.
2 As quoted in Carolyn Strange, 'Reconsidering the "tragic" Scott expedition: cheerful masculine home-making in Antarctica, 1910–1913', *Journal of Social History*, 46:1 (2012), p. 76.
3 Images of Robey in prehistoric costume appear in George Robey, *My Life up till Now* (London: Greening, 1908); and *Looking Back on Life* (London: Constable, 1933). The sketch is treated briefly in Peter Cotes, *George Robey: The Darling of the Halls* (London: Cassell, 1972), pp. 51–2.
4 Photographs of Robey in his prehistoric costume are labelled this way in the catalogue of the National Portrait Gallery, Ax160322, Ax160323, and Ax160324, http://www.npg.org.uk/collections.php (last accessed 18 January 2016).
5 Constance Areson Clark, '"You are here": missing links, chains of being and the language of cartoons', *Isis*, 100:3 (2009), p. 572.
6 See for instance Gillian Beer, *Darwin's Plots: Evolutionary Narrative in Darwin, George Eliot and Nineteenth Century Fiction* (Cambridge: Cambridge University Press, 1983); Gowan Dawson, *Darwin, Literature and Victorian Respectability* (Cambridge: Cambridge University Press, 2007); Nadja Durbach, *Spectacle of Deformity: Freak Shows and Modern British Culture* (Oakland: University of California Press, 2009); Jane R. Goodall, *Performance and Evolution in the Age of Darwin: Out of the Natural Order* (London: Routledge, 2002); Stephanie Moser, *Ancestral Images: The Iconography of Human Origins* (Ithaca: Cornell University Press, 1998); Nicholas Ruddick, *The Fire in the Stone: Prehistoric Fiction from Charles Darwin to Jean M. Auel* (Middletown, Connecticut: Wesleyan University Press, 2009); and Martin J.S. Rudwick, *Scenes from*

Deep Time: Early Pictorial Representations of the Prehistoric World (Chicago: University of Chicago Press, 1992).
7   The quotation is from Bernard Lightman, *Victorian Popularizers of Science: Designing Nature for New Audiences* (Chicago: University of Chicago Press, 2007), p. 14. See also Geoffrey Cantor et al., *Science in the Nineteenth Century Periodical: Reading the Magazine of Nature* (Cambridge: Cambridge University Press, 2004), p. 93; and James G. Paradis, 'Satire and science in Victorian culture', in Bernard Lightman (ed.), *Victorian Science in Context* (Chicago: University of Chicago Press, 1997), pp. 147–8.
8   Alvar Ellegård, *Darwin and the General Reader* (Chicago: University of Chicago Press, 1990), p. 5.
9   'Edward Tennyson Reed', in L.G. Wickham Legg (ed.), *The Dictionary of National Biography, 1931–1940* (London: Oxford University Press, 1949), p. 730.
10 John Lubbock, *Pre-historic Times* (London: Williams and Norgate, 1865), pp. 237–67.
11 See for instance Michael Klossner, *Prehistoric Humans in Film and Television: 581 Dramas, Comedies and Documentaries, 1905–2004* (Jefferson, North Carolina: McFarland, 2006).

1

Mass culture: the Victorian world picture

This very British story of how the Victorians introduced the modern cave man character to global popular culture requires a quick contextual overview of the mid-nineteenth-century intellectual, technological, social and cultural factors through which ideas about prehistory and human evolution reached a broad public. The following is not exhaustive. It distils, synthesises, generalises and melds important concepts from elite and mass culture in ways that are analogous to what the British public once did. My purpose is to equip readers with sufficient knowledge to understand my arguments, while also showing how Victorians constantly encountered, navigated and understood a barrage of ideas from the evidence-based to the blatantly deceitful. Their reactions ranged from wholesale adoption to condemnation by way of mocking humour.

The story begins overseas and long ago among the Greek and Arab scientists and philosophers who examined sea creatures, plants, physical phenomena and celestial bodies, identifying what they believed were divinely created laws governing the origins and evolution of life. Their writings were largely unquestioned until the seventeenth-century emergence of the 'scientific method', which stressed empirical observation. The new approach was famously employed by the eighteenth-century French scientists who were spurred by Enlightenment encyclopaedic impulses and roiling republicanism to uncover observable laws governing the earth. The foremost researchers congregated at Paris's Jardin des Plantes, the national botanical garden, and its neighbour, the Musée National d'Histoire Naturelle, whose vast collections included specimens imported from throughout the world. Scientists and students from many countries burnished their knowledge in Paris.[1]

At the same time, scientists were perplexed by the huge mammal bones that French explorers began discovering in the Ohio River Valley in the 1730s. These creatures closely resembled modern elephants, animals not found in the Americas. The bones sparked the interest of men like the Francophile patriot and polymath Thomas Jefferson, who believed that the creatures epitomised North America's physical grandeur and the colonies' political destiny, and suggested that the animals might still roam the continent's unexplored vastness. Jefferson's enthusiasm was matched by the portraitist, naturalist and collector Charles Willson Peale, who opened the country's first public museum in 1801 in Philadelphia. Among the soberly arranged collections of stuffed animals and plants were two almost complete 'mammoths' from upstate New York: the first fossil skeletons mounted in North America and possibly the world. French scientists had recently used European remains to prove that the animal was not related to the elephant: the first conclusive evidence that a species had become extinct. Peale understood the public interest and advertised the mammoth as the largest creature ever to have walked the earth. The more precise scientific taxonomy 'mastodon' would be coined only in 1806, by which time Americans were applying its colloquial equivalent to such oversized things as loaves of 'mammoth bread' produced by Philadelphia bakers and a 'mammoth cheese' delivered to President Jefferson.[2] Hoping to wring as much profit as possible, Peale sent his sons Rembrandt and Rubens – their brothers were named Raphaelle and Titian – on tour with one of the skeletons. Their departure was feted with a 'mammoth dinner' in February 1802 that was served on a table beneath the skeleton. The brothers displayed the beast in London that September, extolling its size, aggressiveness and carnivorous appetite in the press. But the British public did not warm to an animal associated with American republicanism and French science.[3]

The muted response showed that, radicals and reformers aside, the British greeted Enlightenment scientific ideas with a scepticism bordering on horror. In the eighteenth century, those who speculated about the earth's age like the Derbyshire doctor Erasmus Darwin – grandfather to Charles, of whom more later – were derided in the press, lampooned by writers, and lived in fear of arrest. British science was comparatively conservative. Most practitioners had been taught by ordained dons at Oxford and Cambridge, many of whom adhered to Natural Theology, a set of ideas most famously associated with the Anglican cleric William Paley's argument that just as the intricate mechanism of a pocket watch strongly suggests that it has been deliberately designed, the earth's complexity points to a divine creator. This is not to argue that British science was monolithic, since Edinburgh and London universities offered less theologically inspired training, radical newspapers published favourable reports about evolution, and theologians began attempting to reconcile science and religion by critically examining Biblical texts.[4]

Mastodons were North American and Continental mammals that had clearly coexisted with humans in the fairly recent past. British minds were far more troubled by the fossils of giant lizard-like creatures found within their own shores, most notably near Lyme Regis, Dorset. Locals began collecting and selling fossils to the tourists who flocked to Britain's coasts in the early nineteenth century, having been prevented from crossing the Channel to France by the Napoleonic wars. A growing band of scientifically inclined visitors recognised the fossils as evidence that the planet's deep past might not accord with the Christian belief that the earth was only about 6,000 years old, a figure arrived at by meticulously calculating the duration of Biblical events. In 1811–12, the complete skeleton of a huge animal was discovered at Lyme Regis, exhibited in Piccadilly and subsequently purchased by the British Museum. Anatomists debated about what it was before settling on the equivocal name *ichthyosaurus*, or 'fish lizard'.[5]

Visitors to Lyme Regis included William Buckland, a fellow of Christ Church, Oxford, and a pioneering geologist. His faith was severely tested in 1821 when he examined a cave near Kirkdale in north Yorkshire containing the ancient bones of animals that were no longer found in Britain. Buckland was transformed by the discovery. Every copy of his report, in which he argued that the animals had once been native to Britain and had not been, as some believed, washed into the cave by Noah's flood, was purchased immediately. Other, less prominent scientists confronted similarly uncomfortable truths about Britain's ancient flora, fauna and geology.[6] Many of them began questioning Biblical orthodoxy. Richard Owen, a brilliant anatomist and devout Anglican who headed the British Museum's Natural History department, was less pliant, and so he embodies a conservative reaction. He had attended the University of Edinburgh and witnessed the 1830 revolution while studying in Paris, becoming convinced that 'French' evolutionary ideas were inimical to social order. He made extensive studies of British fossils, deliberately downplaying characteristics that would support their evolutionary importance, and gave this ever-expanding class of extremely ancient and long-lasting reptiles the name 'dinosaur', or 'terrible lizard', in 1841.[7]

Ancient mammals and reptiles were one thing, but the remains of physiologically distinct 'archaic humans' that scientists began unearthing in Britain and Europe in the 1830s prompted further uneasy questioning of the Biblical narrative. When a partial hominid skeleton was discovered in Germany's Neander Valley, or *Tal*, in 1856 it was clear that this was a distinct species, subsequently named *Homo Neanderthal*. Because fossil bones could not be dated accurately until the 1940s, sceptics claimed that such finds belonged to wild tribes described in Roman histories or to mythological monsters. Hominid remains found alongside those of extinct animals were dismissed as more recent graves that had been inadvertently dug into older layers. Then in 1858 a cave was discovered near Brixham, Devon, containing the bones of extinct animals and flint tools. Local amateurs and London experts excavated it methodically, and presented their results to the Society of Antiquaries, one of the nation's most august scientific bodies, in June 1859. After this, 'no British geologist publicly disputed the idea that humans and extinct animals had once coexisted', a consensus that spawned countless lectures and over eighty articles in periodicals aimed at scientific, religious and general readers.[8]

Knowledge of early humans expanded in 1868 with the discovery of fossils in south-western France in a rock shelter or *cro* on the land of a family named Magnon. Further discoveries of Cro-Magnons – the new species name was derived from the place where it was first identified – were made over the next fifty years. Cro-Magnon tools, cave art and small sculptures pointed to intelligence and social sophistication. Many scientists believed Cro-Magnons were modern humans who, it seemed increasingly clear, had appeared on earth far before the Bible suggested they had. Some Christians clung to scriptural narrative, but the famously devout Prime Minister William Ewart Gladstone was not among them. He told a packed meeting in 1872, which had gathered to hear about the discovery of a pre-Biblical account of the Flood, 'I do not know whether it is supposed that inquiries into archaeological and other sciences are to have the effect of unsettling many minds in this our generation, but I must say that for me, as to the very few points on which I am able to examine them, they have a totally different effect.'[9] Gladstone belonged to the Victorian elite, but his eager curiosity about scientific evidence showed an accommodation between faith and reason that crossed class borders. Relatively few Victorians were doctrinaire about either. They understood the tensions between science and religion in some general way that, as we shall see, was exploited by showmen and humorists.

These discoveries coincided with and complemented an increasingly complex understanding of British history. In the early nineteenth century, archaeologists and antiquarians began excavating the stone circles, barrows and hill forts that are such prominent features of Britain's landscape, but that had been largely overlooked in favour of foreign Classical sites. Two pieces of mid-century legislation embodied the growing interest in Britain's past and the belief among religious, political and social elites that an understanding of the nation's history might exert a stabilising influence on a rapidly urbanising, industrial society. The Libraries, Museums and Gymnasiums Act of 1845 led to the creation of dozens of institutions throughout the country in which physical evidence of the nation's past was preserved, studied and displayed. It had been preceded by the Public Record Office Act of 1838, which set the conditions for the emergence of the modern professional discipline of history with its basis in documentary research. The Public Record Office Act centralised authority over dozens of local institutions and authorised the cataloguing of state papers, while the Public Record Office, the repository for this new documentary heritage, opened in London's Chancery Lane in 1851. The Historical Manuscripts Commission was established in 1869 to identify and inventory private collections of historical records. Alongside these developments, the universities introduced courses in modern history, and widely read historians like Thomas Babington Macaulay and Henry Buckle explored the roots of Britain's unwritten, inherited constitution. They expounded a broadly nationalist, teleological progression from immemorial Anglo-Saxon institutions to the pinnacle of Victorian society. This interest in the country's past was more romantically displayed in such things as Sir Walter Scott's historical novels, Pre-Raphaelite paintings, Lord Tennyson's Arthurian verses, Gothic architecture and the gentlemanly ideals inculcated in public schools. These presented a conservative, idealised vision of an ordered, stable, rural Britain that contrasted with the grimy, urban, industrial present.[10]

Britain's geology and history were increasingly well understood, but nineteenth-century maps abounded with blank spaces where ancient creatures like mammoths and primitive hominids might still live. The latter idea was fed by Europeans' familiarity with small monkeys and apes, which had frequently been depicted in mediaeval and early modern art, manuscripts and architecture undertaking human tasks that suggested a tenuous kinship. Great apes, which lack tails and approximate humans in size, were far rarer and considerably more troubling. In the seventeenth century Europeans first encountered chimpanzees in Africa and orang-utans in the Far East, while scientists tried to make sense of the creatures. The Swedish botanist Carolus Linnaeus grouped monkeys, apes and humans together in his influential eighteenth-century taxonomy *Systema Naturae*, and at about the same time the eccentric Scottish peer Lord Monboddo argued more provocatively that orang-utans were human. Owen refuted any relationship just as emphatically a century later. The ideas advanced by these and other scientists became widely known through translations and through reports in British magazines and newspapers. Nonetheless, very few people had ever encountered great apes in the wild. The creatures could only be imagined from written details and drawings until the 1830s, when live examples arrived in London.[11]

Showmen exploited the public fascination with simians as early as 1699, when a visitor to Bartholomew Fair, the raucous late-August gathering that had been held outside London's Aldersgate since the twelfth century, described seeing 'monkeys imitating men, and men mimicking monkeys'.[12] In 1817, Thomas Love Peacock published *Melincourt*, a novel about an English aristocrat who introduces an orang-utan into society as a young man named Sir Oran Haut-Ton and eventually secures him a rotten borough. Sir Oran does not speak, but his Frenchified name, fine tailoring and manners cloak his identity, allowing Peacock to satirise the Romantic idealisation of nature, manners, courtship, evolutionary theories and the unreformed Parliament.[13] The humour played on the contradictions between the human status that Monboddo had awarded the orang-utan and its role as a byword for brutish behaviour. For instance, the poet Samuel Taylor Coleridge dismissed atheism as an 'ouran outang theology', while in 1830 the budding scientist Charles Lyell compared the rebels in the streets of Paris to the animal.[14] The most resonant reaction to questions about the relationship between apes and humans was the play *Jocko*, which opened in Paris in 1825. The story about a monkey that saves a shipwrecked family was little more than an excuse for the acrobat playing the title role to gambol about, but the ape's heroism suggested that it was almost human. The play and its many imitations were popular in Britain and America for forty years.[15] The engraver Thomas Landseer responded to

Jocko's success with *Monkeyana*, a collection of images depicting monkeys dressed as humans accompanied by witty rhymes and quotations from authors like Juvenal, Shakespeare and Jonathan Swift.[16]

The name Jocko was derived from the early seventeenth-century account of Andrew Battel, an English sailor imprisoned by the Portuguese in present-day Angola. He described two creatures – almost certainly gorillas – whose indigenous names he transcribed as Pongo and Engeco. The few Europeans and North Americans who had since travelled to West Africa – a collection of sailors, traders, slavers and missionaries – had been hemmed on the coast by dense jungle, mangrove-clogged rivers and endemic disease. This changed in the 1840s as prophylactic drugs made it safer to venture inland. In 1847, the gorilla was described scientifically in America on the basis of the examination of a partial skeleton. As we might now anticipate, Richard Owen carried out his own study, concluding that the animal bore only passing similarities to humans.[17]

The equanimity with which these scientific discoveries and debates took place reflected the stability of mid-nineteenth-century Britain. The country had weathered economic depression, riots, corn laws, poor laws and the Great Reform Bill, while Chartism's peaceful disintegration in 1848 contrasted with the revolutions that convulsed the Continent. London was the world's largest city, the cultural centre of the English-speaking world and an emerging imperial metropolis. Confidence, material prosperity and political stability were evident in a fetish for innovation and progress. The public's idols included the men who harnessed steam for industrial purposes and spanned Britain with railways: over 7,000 miles of track were laid between 1830 and 1850, effectively integrating the provinces with the capital for the first time. The introduction and rampant spread of trains, steamships, pillar post boxes and railway book stalls throughout Britain symbolised how goods, printed matter, private letters and people moved further and faster than ever before. The Crystal Palace, a vast steel and glass structure erected in Hyde Park for the Great Exhibition of 1851, symbolised the nation's ingenuity and industrial might.

Scientists were public figures in part because the modern profession with its arcane language and abstruse subjects had not fully emerged. Public lectures were regularly hosted at venues as diverse as the Royal Institution, home of the nation's foremost scientific body, and the Royal Polytechnic, an entertainment complex that offered rides in a diving bell alongside the chemist John Henry Pepper's optical illusions.[18] From 1831 onwards hobbyists and professionals gathered at the annual late-summer meeting of the British Association for the Advancement of Science. Middle-class adepts tramped through the countryside, hacking specimens out of rocks, netting all manner of creatures, and investigating archaeological remains. Their passions were satirised in Wilkie Collins's 1868 novel *The Moonstone*, whose narrator, the butler of a Yorkshire country house, observes bemusedly: 'gentlefolks in general have a very awkward rock ahead in life – the rock ahead of their own idleness … it is curious to see – especially when their tastes are of what is called the intellectual sort – how often they drift blindfold into some nasty pursuit', referring to activities like collecting.[19] Likeminded people founded scientific societies throughout the country, in turn creating loose national networks that supplied specimens to London-based specialists. The increasingly structured divisions between work and leisure in urban areas also freed labourers and artisans to pursue intellectual interests. Scientists and social reformers helped diffuse elite ideas by lecturing at mechanics' institutes and working men's colleges; in 1860 there were about 1,200 of these, with some 200,000 members throughout Britain. Less formal, male-dominated working-class groups also provided opportunities for socialising and learning.[20]

Dramatic rises in literacy help to account for the quadrupling in the number of books and magazines published in Britain between 1800 and 1860, though James Secord noted the irony that these were typeset, printed, bound and shipped by workers who earned too little to buy their own copies. However, low wages did not exclude workers from access to books, which they borrowed from mechanics' institutes, shared, bought on market stalls and learned about indirectly through lectures. Middle-class access to books was greatly facilitated by commercial circulating libraries like the one founded in 1843 by the Bloomsbury stationer Thomas Mudie. For a one-guinea annual fee, roughly the cost of a book in mid-Victorian Britain, members could choose from a stock of approximately one million volumes. Mudie's was the very model of a modern Victorian business, with vast stocks of the most popular titles, a fleet of wagons to deliver them throughout the city, and railways and steamships to reach members in the provinces and overseas.[21]

British publishers had first responded to public interest in science in the 1820s by producing heavily illustrated texts. The most successful such book was *Vestiges of the Natural History of Creation* of 1844. Its anonymous author wove a novelistic narrative, Enlightenment ideas and popular science into an all-encompassing evolutionary theory that attracted readers of every political, religious and scientific stripe and propelled discussions about evolution into polite society. Middle-class readers bought or borrowed *Vestiges*, while the working classes mostly learned about it through lectures, discussions and incidental references, causing an immediate, widespread shared knowledge about *Vestiges* that Secord termed a 'sensation'.[22] The book was satirised by Benjamin Disraeli in his 1847 novel *Tancred*, which includes a young woman's attempt to coherently explain a popular scientific treatise entitled *Revelations of Chaos*: 'all is development. The principle is perpetually going on. First, there was nothing, then there was something; then, I forget the next, I think there were shells, then fishes; then we came, let me see, did we come next? Never mind that; we came at last.'[23] The speaker's hesitant grasp of the book fairly accurately reflected many people's vague, generalised knowledge about the arguments made in *Vestiges*. Nonetheless, the book's success spurred British publishers to produce introductory mass-market science texts for decades to come.[24]

Publishing and literacy were in the ascendant, but mid-Victorian Britain was also an intensely visual era. City walls, railings and hoardings were bedecked with brightly coloured advertisements, plays used vivid backdrops, and heavily illustrated newspapers and magazines proliferated thanks to new printing techniques, mechanically produced paper and the abolition of traditional taxes. These publications, which appeared from weekly to monthly and cost as little as a penny, were the core of an emerging 'mass culture' centred in images rather than text, which, as Patricia Anderson convincingly argued, broke down the barriers based on literacy that had traditionally segregated culture along class and socio-economic lines. The terms 'mass' and 'popular' are used interchangeably in this book to refer to this form of broadly inclusive culture. Innumerable magazines, many of them short-lived, were also founded that encompassed diverse views and 'prospered to the extent that they could interpret, or represent, or shape their readers' opinions'.[25] Articles on scientific subjects abounded, mostly being written by generalists who bowdlerised and simplified arcane and abstract ideas, leading Peter Broks to conclude that 'our appreciation of popular science is incomplete without locating it within the popular culture of which it is a part'.[26]

Scholars charting the spread of ideas have estimated that Victorian magazines and newspapers reached five times the numbers indicated by circulation figures, as copies were shared, sold to market traders or pulled apart so that their illustrations could be used to decorate public houses and modest homes.[27] These estimates now seem extraordinarily modest. The full-text search capabilities of digitised newspaper and magazine databases reveal the huge scale of re-publication throughout the country, the empire and the globe. Journalists sold stories repeatedly in different markets, while editors facing the age-old pressure to fill column inches simply lifted content from other publications. Mundane items often appeared dozens of times within a fortnight, a figure that must be multiplied for reports about popular sensations. Sources of such material were sometimes acknowledged, but attributions were not always clear in this pre-copyright era. Constant re-publication spread knowledge of events in London and other centres with astonishing speed to the remotest regions of Britain and beyond.

The era's most successful and comprehensively studied magazine was *Punch, or the London Charivari*, which first appeared in 1841. By the 1860s it had a weekly circulation of 60,000, three times that of its closest rival, though only about a third of the number of copies that *The Times*, the era's most widely read newspaper, sold each day.[28] The comparison between the two publications is apt, since one of *Punch*'s early editors claimed that the magazine set 'its watch by the clock of *The Times*', a judgement averred by its historian, who called it an 'illustrated comic supplement' to the newspaper.[29] *Punch* greatly popularised the concept of 'satire' in which virtually every belief and institution was ridiculed, often through a new visual convention named in the early 1860s a 'cartoon', a word that had traditionally denoted a pencil drawing created to plan a painting, mural or fresco.[30] It was a fitting term for black and white images that guyed topical subjects through visual cues, and contrasted with the colourful, scatological and sexual imagery, and lengthy texts that predominated in Georgian caricatures. The tone also reflected *Punch*'s desire to prod the establishment, while keeping to what the magazine's founding editor called 'the gentlemanly view of things'.[31] *Punch* was a watchword for the middle class; its contributors belonged to this section of society, as did its readers, who lined their libraries with bound editions of the magazine.[32]

Historians have uncovered a great deal about *Punch*'s international reach. The London edition could be purchased in New York and Boston within a week of publication and in the Australian colonies within a month, while cartoons and items from its pages were regularly reprinted in magazines and newspapers throughout Britain, the empire and the United States. Imitative publications, often appropriating the name *Punch*, appeared in the provinces, the United States and virtually everywhere the British settled, from Canada to South Africa and Japan. In the colonies, subscribing to *Punch* or one of its local imitators was a means of asserting a 'British' identity, even as cartoons in the latter focussed increasingly on local subjects.[33]

Punch's tone also shows how popular culture became ever more decorous over the course of the nineteenth century. Nothing better represents this than the fate of Bartholomew Fair, which had long included displays of humans and animals with abnormal bodies. Showmen advertised them as 'freaks', 'monsters' or 'non-descripts' because they did not conform to popular notions of normalcy, while such terms also suggested that they could not be placed within conventional scientific taxonomies. Showmen increasingly incorporated evolutionary concepts in the stories they wove around exhibits, making freak shows an important place in which elite ideas were popularised. But such displays were increasingly anachronistic and attacked by moralists. Freaks were banned from Bartholomew Fair in 1839, and the fair was supressed entirely sixteen years later. By that point, freaks could be seen fairly regularly in the supposedly scientific institutions that dotted London. These venues were also the sites of 'ethnological show business', or exhibits of indigenous peoples, which dated to the earliest days of Britain's global expansion when explorers had brought back and exhibited human specimens. Saddiah Qureshi has shown that the stories that 'experts', be they scientists or showmen, told about indigenous peoples fundamentally shaped how audiences interpreted what they saw. The increasing use of pseudo-scientific language and evolutionary ideas to describe freaks and indigenous peoples testified to what Richard Altick called an 'openly and aggressively displayed aspect of the English character, its complacent assumption of racial supremacy'.[34] Exhibitions in which indigenous peoples were presented as inherently primitive, ignorant and savage also played into the idea of British exceptionalism and helped to justify imperial expansion on the basis that such peoples had to be civilised.

Promoting freaks through inappropriate appeals to scientific authority was perfected by the most influential showman of all time, Phineas Taylor 'P.T.' Barnum. Childhood experience behind the counter of his family's Connecticut shop had immersed Barnum in a Yankee commercial culture that valorised sharp practices like overselling the merits of goods. The historian James Cook found these traits so central to modern American popular culture that he dated its birth to the day in July 1835 when the twenty-five-year-old Barnum began exhibiting Joice Heth, a wizened black woman who, according to the story Barnum wove around her, was 160 years old and had once been the wet nurse to George Washington, the first president of the United States. Heth soon died, but in 1841 Barnum purchased a museum on Broadway in Manhattan, draped it in garish posters and packed it with all manner of exotic animals and freaks. He then unleashed a deluge of outlandish and shameless invocations of scientific theory and popular myth on the pretext that visitors did not have to believe him, but only to feel that they had been entertained. Barnum expounded on the approach he called 'humbug' in lectures and a popular autobiography, thereby changing the word's meaning from deliberate fraud to the stretching of the truth in entertainment.[35]

Barnum first travelled to London in 1844 to exhibit the dwarf performer Colonel Tom Thumb, arriving amid the sensation about *Vestiges*. Sensing an opportunity, he purchased an orang-utan, which he heralded on his return to Manhattan as the 'grand connecting link between the two great families, the human and brute creation' in a fairly typical bit of evolutionary humbug.[36] The creature then inspired Barnum to create 'The wild man of the prairies; or what is it?', the freak he brought to London in 1846 and advertised in explicitly evolutionary terms, by inviting audiences to decide whether this was the 'LINK between Man and the OURANG-OUTANG, which Naturalists have for years decided does exist, but which has as yet been undiscovered?' until, so Barnum claimed, his agents captured one in California.[37] The creature was depicted on posters as an extremely hirsute hominid in a long jerkin and thick pantaloons, walking with a staff, a sign that it could not easily stand upright. It was in fact Hervey Leech, a New York performer with stunted legs who had mimicked walking on all fours in a succession of monkey parts. To play 'What Is It?' Leech stained his hands and feet, donned a hairy costume and roamed a cage grunting and devouring scraps of raw meat. A British acquaintance of Leech's almost immediately exposed the deceit without denting the popularity of 'What Is It?'[38]

Barnum's humbug for 'What Is It?' was echoed over the following years by American and British showmen, who quoted scientists to lend authoritative credence to their exhibitions, while aligning themselves with the public as curious everymen who wanted only to make up their own minds about controversial subjects. The apparent seriousness of such claims drew the attention of *Punch*'s satirists.[39] Such freaks included the 'Aztec

Lilliputians', a microcephalic boy and girl first exhibited in New York City as the last survivors of a caste of demi-gods from a long-lost city in the Mexican jungle. Scientists immediately disproved the story, but the pair were trumpeted in Britain in 1853 with Barnumesque hyperbole as 'living links between the most remote ages and the present'.[40] Richard Owen – omnipresent when an anti-evolutionary opinion was needed – examined them and declared they were humans, but the royal family and thousands of others were fascinated, while the magazine *Household Words* wondered sniffily, 'with whom lies the offence of humbugging the public?'[41] The Aztecs toured Britain and the Continent for years.

Sober connections between science and showmanship were best exemplified when the Crystal Palace was reassembled atop a south London hill in 1854. It was situated in an immense park that included thirty-one giant concrete dinosaurs arranged amid a series of ponds that evoked 'prehistory', a term that had just been coined. The animals were sculpted by Benjamin Waterhouse Hawkins, under Owen's guidance. Hawkins exaggerated the creatures' size to make them easy to see, and to create a sense of awe and fear. The iguanodon was so big that on New Year's Eve 1854 twenty-one scientists, businessmen, inventors and artists dined at a table laid in its unfinished belly, much as the Peales had done with their mammoth in Philadelphia fifty years earlier. Drawings show that it was a boisterous affair, with a platoon of servants advancing on the table carrying bottles and well-charged glasses, and reports of the lubricated diners singing 'the jolly old beast / is not deceased / there's life in him again!'[42] Over one million people saw the dinosaurs in the first year, causing Owen's chief scientific rival, John Edward Gray of the British Museum, to dismiss them as a 'crowning humbug' that succeeded in 'out Barnuming Barnum', and *Punch* to depict them as the denizens of a child's nightmare.[43]

Debunking evolutionary entertainments was like howling at the wind of a rapidly expanding and commercialising popular culture. The Theatre Regulation Act of 1843 had limited the Lord Chamberlain's power to censor plays and permitted the licensing of new theatres. These in turn had fostered the development of extravaganza and burlesque, middle-class styles of performance that revelled in sophisticated 'puns [and] parodies both musical and literary, topical allusions and contemporary slang' that were often aimed at reconciling audiences to social changes.[44] The Act also paved the way for music hall, which soon dominated popular stage entertainment. By 1870, over 400 halls had been erected throughout

Britain, ranging from 'penny gaffs' in the slums to sumptuously appointed and pretentiously named palaces, palladiums, odeons and empires. The raucous, alcohol-sodden mid-century atmosphere gradually gave way to the middle-class respectability of 'variety' entertainment run by heavily capitalised national syndicates. Romantic notions that music hall was a true expression of working-class British character came under critical scrutiny in the 1970s. It has proved difficult to determine what, if any, consistent political or social message patrons took away, though a formidable body of scholarship has traced the interplay of working-class identity, middle-class corporatism, moral reform, radical politics, regionality, gender, American minstrelsy and imperial swagger in the halls.[45]

Music halls worked on a system in which audiences watched a long roster of artistes who each performed a ten- or fifteen-minute 'turn'. An evening at the halls often encompassed high and low culture, from singers, comedians, jugglers, acrobats and dancers to opera, ballet and short scenes from plays. Performers had so little time on stage that they had to establish their roles instantly, in part by appearing as increasingly standardised, easily recognisable characters. The most famous of these were the garishly dressed swells who regaled audiences with tales of drinking, gambling and carousing; the enticingly clad soubrettes who larded songs with sexual innuendo through knowing winks, wagging fingers and thrusting hips; and those who impersonated popular figures like Ally Sloper, the red-nosed tippler from cartoons who first appeared in 1867 in the London comic magazine *Judy*.[46] Railway excursions and steamship travel may have been out of reach for most people, but performers took imitative versions of London's latest sensations to stages throughout the country, the empire and the globe. By the 1860s, British artistes were appearing on the Continent and the eastern seaboard of the United States. Foreign performers were just as commonly seen on British stages, while regular newspaper reports about the latest sensations in Paris, Berlin and New York created a widespread knowledge about events beyond the nation's shores.

Artistes dressed – both on and off stage – in the extravagant, colourful, and unsubtle costumes that suited their extrovert profession. Arresting attire could only fleetingly capture the attention of a crowd distracted by bantering, eating, gambling or consorting with the prostitutes that moralists espied behind every ornate pillar and plush curtain. Transforming them into an 'audience' that was focussed on the stage was the most important skill for performers to master, and the one that distinguished successful artistes from failures. They did this most often by incorporating humorous, melodramatic or heart-rending allusions to topical people and events, relying, as Peter Bailey has shown, on shared knowledge, however generalised or corrupted, of elite mores and debates.[47] Once captured, audiences were encouraged to sing along with deliberately simple melodies and repetitive choruses, many of which remain familiar: 'Champagne Charlie is my name', 'A little of what you fancy does you good', and 'I do like to be beside the seaside'.

These refrains resonate in part because music hall culture extended far beyond the physical walls of the theatres in which it was ostensibly centred. Music and lyrics to the latest tunes were sold in London and the provinces and shipped throughout the globe, where they were often republished, officially or not. While a performer might reach hundreds of people each evening, official editions of the most popular song sheets sold as many as 80,000 copies in London alone.[48] Popular songs were heard in pubs and on the streets, and were played on the pianos that adorned middle-class parlours. The brightly coloured illustrations on song sheet covers, which generally portrayed the singer in her or his stage costume, were also used as inexpensive decorations for bedsitting rooms and beer parlours, reinforcing the fame of individual performers, codifying costumes and cementing mass culture's intensely visual nature.

Songs were composed, and song sheet covers illustrated, by men working simultaneously for magazines and newspapers, writing plays, decorating theatres and any number of other artistic pursuits. For some it was a diverting pastime, while those who sought their livings solely in London's Bohemia faced a precarious and often threadbare, grub-street existence. Playwrights and songwriters earned a flat fee for their work, while *The Times* was the only publication that paid its contributors an exclusive salary. The borders between stage, page, image, text and song were fluid as members of this creative circle wrung as much profit as possible from successful ideas by reusing them in different media.[49] Audiences moved equally fluidly and casually through mass culture's many forums, expecting to see common topical references as they did so. Individuals chose where and how to engage with mass culture largely according to socio-economic and geographical factors, though topical concerns were represented simultaneously in penny gaffs and magazines as well as shilling books and opulent theatres.

By the mid-nineteenth century, mass culture incorporated authoritative scientific voices like Owen's as well as those of showmen like Barnum whose humbugging claims staked out a more fatuous credibility. Audiences had a general knowledge of evolution, without necessarily endorsing it. They were sceptical, dismissive or casually accepting as the circumstances seemed to warrant. They wanted above all to be entertained, which meant increasingly an evening in the music halls. Mass culture was dynamic and was characterised, as we will see, by a constant need for new, attention-grabbing, topical acts. It emerged as the British public's interest in 'primitive races from the corners of the earth had been abundantly satisfied, and showgoers once more were seeking new sensations, perhaps even new food for thought'.[50] This was provided in late 1859 by the unlikeliest of public figures, the reclusive scientist Charles Darwin, when he published his evolutionary treatise *On the Origin of Species*, one of the most influential books of all time. A recent study of the American creationist movement usefully characterised *Origin*'s reception by the public as 'contingent on a complex interplay of timing, nation and culture'.[51] This chapter has attempted to explain the fertile ground on which the British received *Origin*, because it was the seed from which the modern cave man character grew.

Notes

1 Rebecca Stott, *Darwin's Ghosts: In Search of the First Evolutionists* (London: Bloomsbury, 2012), passim.
2 Paul Semonin, *American Monster: How the Nation's First Prehistoric Creature Became a Symbol of National Identity* (New York: New York University Press, 2000), pp. 111–35; Ellis L. Yochelson, 'Mr Peale and his mammoth museum', *Proceedings of the American Philosophical Society*, 136:4 (1992), pp. 493–7; Charles Coleman Sellers, 'Peale's Museum and the "museum idea"', *Proceedings of the American Philosophical Society*, 124:1 (1980), pp. 29–32.
3 John Elihu Hall, 'The mammoth feast', in *The Philadelphia Souvenir* (Philadelphia: William Brown, 1826), pp. 45–55; 'Charles Willson Peale', *Morning Chronicle* (16 March 1802), p. 3; 'Mammoth', *Morning Post and Gazeteer* (8 November 1802), p. 1; 'The great incognitum, or mammoth', *Morning Chronicle* (11 April 1803), p. 1.
4 Bernard Lightman, *Victorian Popularizers of Science: Designing Nature for New Audiences* (Chicago: University of Chicago Press, 2007), pp. 39–94; Stott, *Darwin's Ghosts*, pp. 160–87.
5 Martin J.S. Rudwick, *Worlds before Adam: The Reconstruction of Geohistory in the Age of Reform* (Chicago: University of Chicago Press, 2009), pp. 27–8; A. Bowdoin Van Riper, *Men among the Mammoths: Victorian Science and the Discovery of Human Prehistory* (Chicago: University of Chicago Press, 1993), p. 156.

6 Van Riper, *Men*, pp. 39–73.
7 Adrian Desmond, 'Richard Owen's reaction to transmutation in the 1830s', *British Journal for the History of Science*, 18:1 (1985), pp. 25–50; and Adrian Desmond, 'Richard Owen's response to Robert Edmond Grant', *Isis*, 70:2 (1979), pp. 224–34.
8 The quotation is from Van Riper, *Men*, p. 117; see also pp. 76–116. See also Erik Trinkhaus and Pat Shipman, *The Neanderthals: Changing the Image of Mankind* (New York: Alfred A. Knopf, 1993), pp. 46–51 and 98–106; and Stephanie Moser, *Ancestral Images: The Iconography of Human Origins* (Ithaca: Cornell University Press, 1998), pp. 21–38.
9 'Chaldean history of the Deluge', *The Times* (4 December 1872), p. 7.
10 Barry Cunliffe, 'Introduction: the public face of the past', in John D. Evans, Barry Cunliffe and Colin Renfrew (eds.), *Antiquity and Man: Essays in Honour of Glyn Daniel* (London: Thames and Hudson, 1981), p. 192; Philippa Levine, *The Amateur and the Professional: Antiquarians, Historians and Archaeologists in Victorian England, 1838–1886* (Cambridge: Cambridge University Press, 1986), passim; Jeffrey Richards, *The Ancient World on the Victorian and Edwardian Stage* (London: Palgrave Macmillan, 2009), pp. 1–25.
11 See for instance Stefaan Blancke, 'Lord Monboddo's ourang-outang and the origin and progress of language', in Marco Pina and Nathalie Gontier (eds.), *The Evolution of Social Communication in Primates* (New York: Springer, 2014), pp. 31–43; Thomas Henry Huxley, *Evidence as to Man's Place in Nature* (New York: Appleton, 1863), pp. 17–31; H.W. Janson, *Apes and Ape Lore in the Middle Ages and the Renaissance* (London: Warburg Institute, 1952); and 'History of the orang-outangs', *Scots Magazine* (1 August 1781), pp. 417–21.
12 Henry Morley, *Memoirs of Bartholomew Fair* (London: Chapman and Hall, 1859), p. 350.
13 Thomas Love Peacock, *Melincourt: or Sir Oran Haut-Ton* (London: Macmillan, 1896).
14 The reference to Coleridge is in Adrian Desmond, 'Artisan resistance and evolution in Britain, 1819–1848', *Osiris*, 3 (1987), p. 108. The reference to Lyell is in Stott, *Darwin's Ghosts*, p. 238.
15 Jane R. Goodall, *Performance and Evolution in the Age of Darwin: Out of the Natural Order* (London: Routledge, 2002), pp. 50–2; Diana Snigurowicz, 'Sex, simians and spectacle in nineteenth century France: or how to tell a man from a monkey', *Canadian Journal of History*, 34 (1999), p. 52; 'American theatricals', *Era* (7 September 1862), p. 10; 'Pantomimists', *Era* (24 January 1864), p. 16.
16 Thomas Landseer, *Monkeyana, or Men in Miniature* (London: Moon, Boys and Graves, 1827).
17 Colin Groves, 'A history of gorilla taxonomy', in Andrea B. Taylor and Michele L. Goldsmith (eds.), *Gorilla Biology: A Multidisciplinary Perspective* (Cambridge: Cambridge University Press, 2002), pp. 15–18; 'The new man monkey', *Liverpool Mercury* (7 February 1859), p. 3.

18 Lightman, *Victorian Popularizers*, pp. 9–38 and 167–218.
19 Wilkie Collins, *The Moonstone* (London: Tinsley Brothers, 1868), p. 99.
20 Desmond, 'Artisan', pp. 77–110; Adrian Desmond, *Huxley: From Devil's Disciple to Evolution's High Priest* (Reading: Addison Wesley, 1997), pp. 208–11; Alvar Ellegård, *Darwin and the General Reader* (Chicago: University of Chicago Press, 1990), pp. 63–7; John Laurent, 'Science, society and politics in late nineteenth-century England: a further look at mechanics' institutes', *Social Studies of Science*, 14:4 (1984), pp. 585–619; Lightman, *Victorian Popularizers*, pp. 197–212; Anne Secord, 'Science in the pub: artisan botanists in early nineteenth-century Lancashire', *History of Science*, 32 (1994), pp. 269–315.
21 James A. Secord, *Victorian Sensation: The Extraordinary Publication, Reception and Secret Authorship of 'Vestiges of the Natural History of Creation'* (Chicago: University of Chicago Press, 2000), pp. 304–8; Guinevere L. Griest, 'A Victorian leviathan: Mudie's Select Library', *Nineteenth-Century Fiction*, 20:2 (1965), pp. 103–26.
22 Secord, *Victorian Sensation*, pp. 9–40 and 155–90.
23 Benjamin Disraeli, *Tancred, or the New Crusade* (London: Henry Colburn, 1847), p. 130.
24 Lightman, *Victorian Popularizers*, pp. 19–33.
25 The quotation is from Alvar Ellegård, 'Public opinion and the press: reactions to Darwinism', *Journal of the History of Ideas*, 19:3 (1958), p. 384. See also Patricia Anderson, *The Printed Image and the Transformation of Popular Culture, 1790–1860* (Oxford: Clarendon Press, 1991), passim; Alvar Ellegård, 'The readership of the periodical press in mid-Victorian Britain', special issue, *Acta Universitatis Gothoburgensis*, 63:3 (1957), passim; Thomas Richards, *The Commodity Culture of Victorian England: Advertising and Spectacle, 1851–1914* (Stanford: Stanford University Press, 1990), passim; and Richards, *Ancient World*, pp. 28–65.
26 Peter Broks, 'Science, media and culture: British magazines, 1890–1914', *Public Understanding of Science*, 2:2 (1993), p. 136. See also Marianne Sommer, 'Mirror, mirror on the wall: Neanderthal as image and "distortion" in early 20th-century French science and press', *Social Studies of Science*, 36:2 (2006), p. 209.
27 Lucy Brown, 'The treatment of the news in mid-Victorian newspapers', *Transactions of the Royal Historical Society*, 27 (1977), p. 39; Henry J. Miller, 'John Leech and the shaping of the Victorian cartoon: the context of respectability', *Victorian Periodicals Review*, 42:3 (2009), p. 267.
28 Richard Noakes, 'Science in mid-Victorian *Punch*', *Endeavour*, 26:3 (2002), pp. 92–6.
29 Both are as quoted in Richard Noakes, '*Punch* and comic journalism in Victorian Britain', in Geoffrey Cantor et al., *Science in the Nineteenth-Century Periodical: Reading the Magazine of Nature* (Cambridge: Cambridge University Press, 2004), p. 94.
30 Martin Fichman, *Evolutionary Theory and Victorian Culture* (Amherst, New

York: Humanity Books, 2002), pp. 47–50; 'Cartoon', in *The Oxford English Dictionary*, vol. II (Oxford: Clarendon Press, 1978), p. 140.
31 The quotation is from Noakes, '*Punch*', p. 97. See also Janet Browne, 'Charles Darwin as a celebrity', *Science in Context*, 16:1 (2003), p. 183; and Miller, 'John Leech', pp. 267–91.
32 James G. Paradis, 'Satire and science in Victorian culture', in Bernard Lightman (ed.), *Victorian Science in Context* (Chicago: University of Chicago Press, 1997), pp. 147–50.
33 Simon Potter, 'Communication and integration: the British and Dominions press and the British world, c. 1876–1914', *Journal of Imperial and Commonwealth History*, 31:2 (2003), pp. 190–206; Richard Scully, 'A comic empire: the global expansion of *Punch* as a model publication, 1841–1936', *International Journal of Comic Art*, 15:2 (2013), pp. 6–35. See also '*Punch* in India', *All the Year Round* (26 July 1862), pp. 462–9; and '*Punch* in Australia', *All the Year Round* (22 August 1863), p. 610–16.
34 The quotation is from Richard D. Altick, *The Shows of London* (Cambridge, Massachusetts: Harvard University Press, 1978), p. 279. See also Saddiah Qureshi, *Peoples on Parade: Exhibitions, Empire, and Anthropology in Nineteenth-Century Britain* (Chicago: University of Chicago Press, 2011), pp. 2–5; Nadja Durbach, *Spectacle of Deformity: Freak Shows and Modern British Culture* (Oakland: University of California Press, 2009), pp. 2–32 and 90–2; Goodall, *Performance*, pp. 70–1; and Paul Semonin, 'Monsters in the market place: the exhibition of human oddities in early modern England', in Rosemarie Garland Thomson (ed.), *Freakery: Cultural Spectacles of the Extraordinary Body* (New York: New York University Press, 1996), pp. 71–8.
35 James W. Cook, *The Arts of Deception: Playing with Fraud in the Age of Barnum* (Cambridge, Massachusetts: Harvard University Press, 2001), p. 3; Phineas Taylor Barnum, *The Autobiography of P.T. Barnum* (London: Ward and Lock, 1855), pp. 85–6; and Eric Fretz, 'P.T. Barnum's theatrical selfhood and the nineteenth-century culture of exhibition', in Thomson (ed.), *Freakery*, pp. 97–100.
36 John Rickards Betts, 'P.T. Barnum and the popularization of natural history', *Journal of the History of Ideas*, 20:3 (1959), pp. 353–4.
37 Barnum's advertising pamphlet for 'The wild man of the prairies; or what is it?' at the Egyptian Hall, London, probably 1846, National Fairground Archives, University of Sheffield.
38 'Theatre Royal, Drury Lane', *Morning Post* (31 January 1838), p. 2; 'The wild man of the prairies', *The Times* (1 September 1846), p. 6; 'Wild man of the prairies', *Illustrated London News* (5 September 1846), p. 154; 'American swindles', *Era* (15 November 1846), p. 11; 'Death of Hervey Leach' [sic], *Illustrated London News* (20 March 1847), p. 189; and Goodall, *Performance*, p. 52.
39 'The deformito-mania', *Punch* (4 September 1847), p. 90; Durbach, *Spectacle*, pp. 23–32.
40 'The Aztec Lilliputians, or Kaanas of Iximaya', *Liverpool Mercury* (21 June 1853), p. 483.

41 The quotation is from 'Lilliput in London', *Household Words* (13 August 1853), p. 573. See also 'Edinburgh', *Era* (13 November 1853), p. 11; 'The Aztec children', *Manchester Times* (1 February 1854), p. 5; 'The Aztecs before the emperor', *Era* (15 July 1855), p. 2; Robert D. Aguirre, 'Exhibiting degeneracy: the Aztec children and the ruins of race', *Victorian Studies Association of Western Canada*, 29:2 (2003), pp. 44–5; and Thomas 'Whimsical' Walker, *From Sawdust to Windsor Castle* (London: Stanley Paul, 1922), pp. 60–1.

42 W.J.T. Mitchell, *The Last Dinosaur Book: The Life and Times of a Cultural Icon* (Chicago: University of Chicago Press, 1998), p. 97.

43 The quotation is from James A. Secord, 'Monsters at the Crystal Palace', in Soraya de Chadarevian and Nick Hopwood (eds.), *Models: The Third Dimension in Science* (Stanford: Stanford University Press, 2004), p. 158. See also 'The Crystal Palace at Sydenham', *Illustrated London News* (7 January 1854), p. 22; and 'The effects of a hearty dinner after visiting the antediluvian department at the Crystal Palace', *Punch* (3 February 1855), p. 50.

44 The quotation is from Jeffrey Richards, *The Golden Age of Pantomime: Slapstick, Spectacle and Subversion in Victorian England* (London: I.B. Taurus, 2015), p. 3; see also pp. 2–34.

45 Important studies include Peter Bailey, *Leisure and Class in Victorian England: Rational Recreation and the Contest for Control, 1830–1885* (Toronto: University of Toronto Press, 1978); Peter Bailey, (ed.), *Music Hall: The Business of Pleasure* (Milton Keynes: Open University Press, 1986); J.S. Bratton, (ed.), *Music Hall: Performance and Style* (Milton Keynes: Open University Press, 1986); and Dagmar Kift, *The Victorian Music Hall: Culture, Class and Conflict* (Cambridge: Cambridge University Press, 1996).

46 Roger Sabin, 'Ally Sloper on stage', *European Comic Art*, 2:2 (2009), pp. 205–25.

47 Peter Bailey, 'Conspiracies of meaning: music hall and the knowingness of popular culture', *Past & Present*, 144 (1994), pp. 138–70.

48 Ronald Pearsall, *Victorian Sheet Music Covers* (Detroit: Gale, 1972), p. 44.

49 Richard Scully, '"The epitheatrical cartoonist": Matthew Somerville Morgan and the world of theatre, art and journalism in Victorian London', *Journal of Victorian Culture*, 16:3 (2011), pp. 363–84.

50 Altick, *Shows*, p. 287.

51 Jeffrey P. Moran, *American Genesis: The Antievolution Controversies from Scopes to Creation Science* (New York: Oxford University Press, 2012), p. 3.

2

Darwin, Du Chaillu and Mr Gorilla: the lions of the season

Professor Richard Owen, the foremost Victorian anatomist, watched excitedly as a barrel that had arrived from Africa with the embalmed carcass of a gorilla was opened in a London laboratory on 11 September 1858. Owen had published a pioneering study of the animal a decade earlier, but he had never seen a complete specimen, thanks to the difficulties in shipping one from Africa's heat and humidity. We can imagine his dismay when the seal was breached and the room filled with a noxious odour such as the River Thames had emitted that summer when effluent and record temperatures had created what newspapers dubbed the 'Great Stink'. Owen and the other nauseated onlookers reached for their handkerchiefs: ineffective protection against the foul miasma. But Victorian scientists were made of stern stuff. The cask was resealed and deposited in a south London field, where staff from the nearby Crystal Palace undertook the distinctly unpleasant job of unpacking, drying and mounting the remains.[1]

Londoners were so unfamiliar with gorillas that when the animal was displayed at the Crystal Palace at year's end, the *Morning Chronicle* newspaper explained that it 'resembles more nearly than any known creature the members of the human family', and a year later the magazine *Children's Friend* still described it as 'a sort of ape [that] has been only lately discovered'.[2] Showmen subsequently capitalised on the heightened public interest by developing ape sketches like the one performed at the London Alhambra, in which an acrobat in a hairy suit, dubbed the 'Man-monkey, or the Rocky Mountain wonder', vaulted onto a moving horse. Monkey costumes were also ubiquitous that winter at Highbury Barn, a vast north London dance hall.[3] These immediate responses show that the public's interest in the creature incorporated a vague sense that it might be related to humans.

This generalised knowledge became acutely focussed by the publication of Charles Darwin's *On the Origin of Species* in late November 1859. The book, which was aimed at the huge mid-Victorian readership for science, set out the interconnected theories of 'Evolution', which held that all living creatures had changed over aeons, and 'Natural Selection', the undirected process of adaptation to climate and other factors.[4] All 1,250 copies of *Origin* were purchased in a single day and a further 3,000 sold almost as quickly in January. Mudie's Select Library bought 500 copies, greatly expanding the book's middle-class reach.[5] The tally would make most writers salivate, though Alvar Ellegård cautions that few of Britain's thirty million inhabitants could afford the book, learning about it instead through the polemical debates that followed its publication. A generalised ignorance about what exactly Darwin had written and the extremely contentious language in which his ideas were publicly discussed led to them almost immediately being dubbed the 'ape theory', even though the creature had never been mentioned in *Origin*.[6] Darwin had anticipated this reaction and had tried to deflect it by sharing his manuscript with influential men on both sides of the Atlantic. Once the book was published, he directed a small group of well-placed scientists who defended his ideas so vociferously that his name remains virtually synonymous with evolution in the public imagination.[7]

Origin's most redoubtable champion was Thomas Henry Huxley, a thirty-five-year-old brilliant and combative anatomist who famously described himself as Darwin's 'bulldog'. Huxley had left school at the age of ten for a medical apprenticeship, progressing eventually to a degree from University College, London. He had then toured the South Pacific as a naval surgeon, collecting specimens at every port of call. It was a humble background that left Huxley committed to working-class education and set him apart from gentlemen scientists like Darwin, while his religious scepticism – he eventually coined the term 'agnostic' to describe the stance that there is insufficient evidence to prove God's existence – and virulent anti-clerical streak distanced him from ordained Oxbridge scientists and devout practitioners like Owen. The pair had once been close, though by the late 1850s they fought openly, to the embarrassment of many colleagues.[8]

At the 1860 summer meeting of the British Association for the Advancement of Science at Oxford, Owen and Huxley clashed about whether gorillas have a hippocampus, the part of the brain that controls memory and spatial navigation, the existence of which would indicate proximity to humans. Owen, who had dissected more great apes than anyone, maintained that they lacked a hippocampus, while Huxley argued the opposite. A second and far more dramatic showdown took place on 30 June in the open discussion after a seemingly innocuous paper that touched on Darwin. This time, Huxley faced Samuel Wilberforce, the fifty-four-year-old Bishop of Oxford and one of Britain's best-known clerics. His father had been a leader of the anti-slavery movement, while he himself had initiated considerable Church reforms, not least in his own diocese, which was rent by Anglo-Catholicism. Critics called him 'Soapy Sam' for his seemingly evasive responses to vexing questions. He was also interested in science and regularly attended the association's annual meeting, though his family's evangelical stamp meant he upheld Biblical truth. His friend Owen had helped him prepare for the debate.[9]

As many as a thousand people jammed the Natural History Museum's hall, ringed the open windows and spilled out onto the lawn in the sweltering afternoon heat, anticipating a showdown between apparently irreconcilable worldviews. But the debate was as limpid as the weather until Wilberforce tried to skewer his opponent by asking if he was descended from an ape on his mother's or father's side. Huxley's response was difficult to hear and so several slightly different versions appeared in London newspapers in the following days and in New York City shortly thereafter. The lack of a definitive transcription then allowed Huxley's supporters to shape and aggrandise the exchange to the extent that by the end of the century it was generally believed he had said, 'I would rather be descended from an ape than a bishop', in what is popularly remembered as the definitive clash between evolution and religion. A recent work has characterised the mood of the educated public after the debate as 'one of muddled curiosity', but as we will see, a wide-ranging interest crossed the spectrum of British society, without being especially confused.[10]

Huxley then promoted his views from the lecture platforms of working men's colleges and mechanics' institutes, explaining Darwin's ideas and arguing that gorillas and humans were closely related, and equally crucially that apes were no closer to black Africans than to any other humans.[11] His lectures were widely reported, an interest that *Punch* reflected in an article written in the voice of 'one who has been spending half an hour or so with Darwin', as fleeting a relationship with *Origin* as that held by many of the debate's loudest participants. *Punch*'s humorist wondered why so many marriages were 'unnatural selections' that paired short men with tall wives, or an 'ugly lout' with a beautiful woman.[12]

Huxley and Owen's ongoing jousting inspired the cleric, naturalist, novelist and erstwhile Darwinian Charles Kingsley to refashion their animus as a great joke. At the 1862 meeting of the British Association, he gave a paper in the person of Lord Dundreary, the monocled, luxuriantly bewhiskered and entirely witless aristocrat from the popular play *Our American Cousin*. Kingsley invoked Dundreary's proclivity for muddling facts and malapropisms to confuse 'hippocampus' with 'hippopotamus' and reversed Huxley's and Owen's intellectual positions.[13] Reports of Kingsley's spoof may have inspired George Pycroft, a surgeon and amateur naturalist in rural Devon. In 1863 he wrote a squib recounting the trial of street-fighting Cockney barrow boys named Dick Owen and Tom Huxley, who associated with a panoply of scientists, including one named Charlie Darwin.[14] It was a localised evocation of the feud that showed how satire's detached bemusement was an almost ideal voice in which to respond to evidently bitter elite discussions.

The Oxford debate also resonated in America thanks to pre-existing trans-Atlantic interest in evolution. Darwin had tried to control the American reaction to *Origin* by sending copies to sympathetic scientists, most notably the Harvard botanist Asa Gray. Darwin's most prominent detractor was Louis Agassiz, Harvard's professor of zoology and geology, who often invoked racist ideas about the inferiority of blacks, amplifying pre-existing tensions about slavery that would pitch the country into civil war in April 1861.[15]

At virtually the same time as Darwin's ideas arrived in America, Paul Du Chaillu, a slight and nervy twenty-nine-year-old, came to the attention of the country's press. Du Chaillu's father was a French trader in Africa while his mother, whose identify is unknown, may have been black. He had grown up in West Africa, where a missionary had befriended him and eventually secured him a teaching post in New York State. Du Chaillu embraced the country and adopted the American pronunciation of his surname as 'Doo Shayloo'. The Philadelphia Academy of Natural Sciences learned of his background and sent him in 1856 to collect birds in West Africa, from where he returned three years later with hundreds of specimens, including twenty-one gorilla skins, the most tangible proof yet of the animal's appearance. Du Chaillu soon fell out with the Academy and, hoping to profit from his stuffed gorillas, rented premises on Broadway, the heart of the American entertainment industry. He lectured about gorillas to learned bodies, but he was neither a scientist nor an adept showman like P.T. Barnum, whose museum was only a few blocks away.[16]

Barnum responded to Du Chaillu by updating 'What Is It?', the evolutionary specimen he had exhibited in New York and London the 1840s. In its original incarnation, 'What Is It?' had been advertised as a creature from California. Barnum now appropriated Du Chaillu's story by claiming that the beast had been captured in West Africa, and cast a young black man in the role. He wore a hairy costume and acted out a simple evolutionary scenario that was explained by a supposedly learned lecturer. Barnum also played on the acrimonious American debates about slavery, displaying 'WhatIs It?' on stage during intermissions of Dion Boucicault's melodrama *The Octoroon* in the museum's theatre. The play follows a young man who inherits a Louisiana plantation, only to fall in love with one of its inhabitants, a beautiful woman named Zoe who confesses that she is an 'octoroon', a slave-era term identifying a person with one black great-grandparent. She is therefore his property and they cannot marry. By inserting 'What Is It?' into the play, Barnum implicitly linked Zoe's racial taint with the more distant evolutionary stain apparently carried by all humans. A report about events in New York that was published in the *Era*, the main British theatrical newspaper, noted that 'What Is It?' 'is related in some way to the irrepressible nigger, and at the same time remotely connected to the gorilla'.[17] The passage reflected explicitly racist descriptions in the New York press, which all suggested that blacks were animalistic and unevolved. Such associations had not been made in America or Britain when the creature had first been exhibited in the 1840s.[18]

By the start of 1861, Du Chaillu realised that he could make no further headway in America, so he left for London, where he was taken up by Owen and Sir Roderick Murchison, the President of the Royal Geographical Society, who understood the collection's scientific value. Murchison invited Du Chaillu to address the society in late February before an audience that included Owen, the Biblical archaeologist Austen Henry Layard and the Chancellor of the Exchequer William Ewart Gladstone. Du Chaillu was flanked on the podium by stuffed gorillas as he spiced his lecture with tales of African barbarity and imitated the animal's call and lumbering walk. Owen then gave an anatomical description of the animal.[19] Gladstone, at least, was moved, stating to the audience that he 'wished it were in his power more frequently to appear in the character of a pupil'.[20] Though the evening was hosted by a sober and learned body, its structure closely resembled those employed by evolutionary freak shows and exhibitions of indigenous peoples. Du Chaillu and his gorillas were the oddities on display, Owen was the objective scientist who guided the audience in understanding what they saw, and Gladstone represented an educated and devout but open-minded section of the middle class. A month later, Murchison chaired a similar lecture by Du Chaillu at the Royal Institution, which was attended by a coterie of eminent scientists.[21]

In early May, Du Chaillu's book *Explorations and Adventures in Equatorial Africa* was published by John Murray, who had brought out *Origin* eighteen months earlier. Mudie's library ordered 1,000 copies, and eight times that number were sold by mid-summer, even if many scientists, Owen's British Museum rival John Edward Gray chief among them, derided the book as a hodgepodge of inaccuracy and fabrication. Du Chaillu and his ghost writer had invested the story with novelistic touches, but the scientific establishment's very public support meant that methodological lapses, grand claims and adventure-story prose could not be overlooked as Barnumesque humbuggery. Gowan Dawson has also revealed tensions at the book's heart between the illustrations in which male gorillas' genitals were obscured by leaves, and folk tales about their aggressive sexual appetite for women. The primness reflected Western traditions for depicting nudes, while also demonstrating intense discomfort at the implication that at some point in the distant past, gorillas and humans had interbred.[22] The mental image of a male gorilla forcing itself onto a woman played into Victorian notions of female sexual passivity and a smug, chauvinistic disgust at the idea that an English gentleman would be so debased as to copulate with an ape.

Du Chaillu's legitimacy was underlined over the summer, when his modestly draped gorillas were displayed at the Royal Geographical Society and Owen purchased a pair for the British Museum. Criticisms continued nonetheless, peaking during Du Chaillu's early-July lecture to the Ethnological Society when he leapt into the crowd and spat angrily at a critic. Du Chaillu's apology for what he called a 'lapse in gentlemanly behaviour', which appeared in *The Times* a few days later, testifies to the pressure exerted by the rancorous debate.[23]

Amplified public interest in gorillas and their human-like qualities caught the eye of satirists. In mid-May, *Punch* published the anonymous thirteen-stanza poem 'Monkeyana', which has since been attributed to the palaeontologist Sir Philip Egerton. He wrote as the gorilla in the first person singular, a novel anthropomorphic perspective that began shifting portrayals of these supposedly ancient predecessors of modern humans from the realm of nightmares to that of comedy. The gorilla's naïve question 'Am I satyr or Man?' broached the fundamental discomfort about whether it embodied mythic animal lust or belonged to the human family. The light verses disguised a fairly profound knowledge of evolution that *Punch* expected its readers to possess. The illustration above the poem contained even more sophisticated messages. It showed a gorilla standing with a staff in its hand, signifying its evolutionary struggle to walk upright and evoking posters for Barnum's 'What Is It?'. More significantly, the placard around its neck read 'Am I a Man and a Brother?', echoing one of the best-known anti-slavery images, a medallion created in 1787 by the Quaker potter Josiah Wedgwood, who as many people knew was Darwin's maternal grandfather. Readers might also have perceived an allusion to the previous summer's Oxford debate, thanks to the role that Bishop Wilberforce's father had played in the anti-slavery movement. Those who could not fully decode the image still saw the deeply troubling question about the physical kinship between humans and apes at the heart of evolution.[24]

Punch revisited the controversy the following week with 'The lion of the season', a more provocative cartoon, whose title invoked the Victorian slang for 'sensation' and a term the British press had used to describe Du Chaillu.[25] The cartoon depicted a formally attired gorilla arriving at a dinner party. Its elaborate whiskers, clothes, heavy watch chain and fobs, along with the surroundings and uniformed footman, reflect decorous, middle-class prosperity. But the toes projecting through the gorilla's shoes remind viewers that the domestic idyll has been invaded by a wild, carnal creature; the essence of the evolutionary threat. It is a freighted image that could easily have been a scene from popular theatre. The footman, the only person who sees what is happening, looks directly at the viewer, quaking in pigeon-toed fright as he stammers out 'Mr. G-G-G-O-O-O rilla!' Referring to the animal with a gentleman's title further accentuates its humanity, while the footman's expression warns the audience of the farce that will ensue when the gorilla enters the parlour, just as a music hall comedian or pantomime principal boy addressed audiences directly. It was a device that drew audiences into what Peter Bailey has called 'conspiracies of meaning' about shared, generalised understandings of elite ideas on which comedy could be built.[26]

The cartoon propelled the gorilla into the mass imagination. Within a week of its publication, the young roisterers at the Epsom Derby – many of them decked out in their best for a day-long spree – were heard hailing one another as 'Mr Gorilla'.[27] Similar references can be detected throughout the summer, including the letter in a Suffolk newspaper announcing 'that in going from Felixstowe Church, along a narrow road, you presently come to a white gate, and by turning down this road you will see the gorilla at the window of a farm-house on your left'.[28] Only the most credulous would have followed the instructions, though many people queued outside a tent selling one-penny glimpses of 'the great gorilla' at an early-August fete near Bath. They entered to find themselves reflected in a mirror, a rather obvious joke that was taken in good humour, judging by the fact that the tent collected over £1.[29] The idea was replicated by a line of leather cases, resembling the ones used to protect daguerreotype photographs, embossed with the motto 'Portrait of the gorilla, taken from life'. A case could be passed to an unsuspecting friend, who would open it to reveal his or her face reflected in a mirror that sat where the photograph should be.[30] The comic magazine *Fun* echoed the idea later in the year by advising 'country cousins who wish to see the gorilla without the expense of a trip to London. – Buy a looking glass'.[31] Similar turns of phrase echoed over the coming years, like the use of 'Mrs Gorilla' to demean women's physical attractiveness and social aspirations.[32] These were all superficial, light-hearted and thoroughly evasive responses to troubling questions about the relationship between humans and gorillas.

A similar reaction was seen at a matinee at the Lyceum Theatre on 19 June. James Planché, a pioneer of topical pantomime and extravaganza, wrote a pun-filled speech in the gorilla's voice.[33] It was declaimed by *Fun*'s young editor Henry J. Byron, who epitomised London's bohemia, having successively abandoned medical and legal studies for the theatre.[34] He began:

> From a gay woodcut – no dull tract with trees on,
> Behold me here – 'the Lion of the Season,'
> MR GORILLA – I announce myself.
> For the Stage doorkeeper – poor timid elf –
> Soon as he saw me in the distance dim
> Bolted! – no doubt for fear that I should bolt him.
> His fear was groundless. Really I am not
> The Great Gorilla Monsieur Chaillu Shot.

> That monster about whom there's so much jaw,
> Must be the perfect one the world ne'er saw.
> Nor am I e'en like those whose bones you see;
> But debonnaire and full of bonhommie,
> In short, of Mr *Punch's* own creation;
> Proof of his powers of investigation.
>
> To speak the Prologue, why they pitched on me,
> I'm thought a link – though some the fact dispute –
> Between the genus Homo and the Brute …[35]

It was another witty, irreverent take on the current debate, declaimed directly by the gorilla.

This was followed at the Adelphi Theatre on 1 July by *Mr Gorilla*, an updated version of Henry Addison's 1842 ape farce *The Blue-Faced Baboon*. It was an obvious choice for the Adelphi, whose manager, as one newspaper proclaimed, 'has seen fit now and then, to present before his patrons, a trifle hitting at the times … M. Du Chaillu now comes in for his share of the public attention.'[36] The production starred the popular comics J.L. Toole and Paul Bedford. Toole played a dry-salter named Pipkin, whose occupation epitomised mercantile stolidity. He had retired to a suburban villa, the setting depicted in *Punch*'s cartoon, an association that was further underlined by the character's fascination with Du Chaillu's book. The action began with Pipkin banishing his ward's suitor from the house. So the young man convinces a friend, played by Bedford, to impersonate an African explorer who just happens to be touring England with a live gorilla, actually, of course, the suitor dressed like *Punch*'s cartoon. The pair are invited to stay, causing antic chaos as the lovers try to meet in secret. It climaxes with the young man saving Pipkin from poisoning himself, in thanks for which he consents to the marriage. Critics were unimpressed, with one claiming that the sketch 'owes whatever success it achieved to the appropriation of the title', showing how the sensation surrounding *Punch*'s cartoon had primed the public.[37] Nonetheless, *Mr Gorilla* played at the Adelphi until late August after which the company took it to Dublin, Belfast and Edinburgh.[38]

Punch's image was recreated most concretely and influentially in the song 'Mr Gorilla, or the lion of the season', of which Howard Paul, an extrovert thirty-year-old expatriate American, gave the first performance in mid-summer. The lyrics were written by Henry J. Byron, who gave the animal a stew of references to sing. His lovelorn gorilla was another satirical sublimation of the sexual tension surrounding the gorilla. Paul took the stage clad as *Punch*'s animal to intone:

> Good Ladies and Gentlemen how do you do
> I am the gorilla of Monsieur Chaillu
> They say that my features a compromise show
> Between chimpanzee and Billy Barlow
> Oh! Dear this terrible gorilla's the rage of the Season you know …
>
> Come dear ladies here's a chance
> Gracious is there no advance
> This fine creature pray behold
> Just a-going to be sold
> Going for a perfect song
> 'Pon my life it's very wrong
> Such a charming foreign swell's been seldom seen
> Oh lovely claws
> Oh handsome jaws
> A complexion quite perfection
> See my carriage
> Who's for marriage
> Hey brown tho' very, very brown
> Better's a brunette than an Albino …
>
> They say I'm a link
> I don't think it's true
> 'Twixt man a brute creation
> Tho' I've no doubt it's the result
> Of deep investigation …[39]

The simple sketch that Paul built around the song caused one critic to sniff that 'it is scarcely likely that the character will have more than an evanescent success, inasmuch as the gorilla has at best a far from charming appearance'.[40] He could not have been more wrong, because Paul performed the song throughout Britain and Ireland for months, promoting his appearances with posters bearing the image that Alfred Concanen, one of the era's most prolific illustrators, had drawn for the sheet music. This showed *Punch*'s gorilla entering a parlour with opened-armed swagger; the scene that followed on from the one that had appeared in the magazine.[41]

The song's popularity inspired several imitations, most notably 'The gorilla quadrille', which was first performed at London's Highbury Barn dance hall in November. The venue's bandleader parodied the tensions between civility and animal brutishness by composing a tune that paired the beast with a jaunty and elegant dance that he conducted while dressed as *Punch*'s gorilla, a scene that Concanen reproduced for the sheet music. There are no reports of dancers taking the floor in the monkey costumes they had worn at the Barn two years earlier, or of whether anyone sang the simplistically racist lyrics that went 'My name it is gorilla and by that you plainly see. By birth I am a Darkie but you can't get hold of me.' Revellers swayed to the song nightly over the winter, during which time other satirical poems and letters appeared that were also written in the voice of Mr Gorilla.[42] Sheet music for these gorilla songs was republished in the United States.[43]

Du Chaillu helped to foster this ongoing interest by lecturing to popular and professional audiences throughout Britain. His September 1861 appearance at the British Association for the Advancement of Science gives a sense of his flair. After being introduced by Owen, he delivered a worthy geographical lecture before pausing dramatically and excusing himself, it is satisfying to imagine, with a knowing inflection and a conspiratorial wink, for not having mentioned the gorilla. He then pointed to a large reproduction of the most famous and contentious image from his book, which had been draped from the balcony. He admitted that the artist had aggrandised the animal for dramatic effect, and gave a sophisticated anatomical description before adapting *Punch*'s colloquialism to claim that 'Madame Gorilla' was apt to run away from humans.[44] His description of female gorillas as inherently bashful further underlined the fear that humans and apes could be related only if at some point in the distant past a male gorilla had raped a woman. Du Chaillu repeated the performance – an apt description for something less than a scientific lecture but more than a music hall turn – throughout the year.

Benjamin Waterhouse Hawkins, who had sculpted the Crystal Palace dinosaurs in the 1850s, also lectured about gorillas in music halls and before sober scientific bodies. At each stop he decorated the stage with images of gorillas and a banner bearing *Punch*'s satirical question, 'Am I not a man and a brother?' Hawkins sketched chalk images as he gave a talk that alluded to Du Chaillu, compared gorillas to Australian aborigines – believed by many Victorians to be the least evolved modern humans – and evoked the French high-wire walker Charles Blondin to distinguish human dexterity, grace and equilibrium from apish strength.[45]

Contrasting gorillas with Blondin was a nod to the latter's fame. He had repeatedly crossed the Niagara Gorge in 1859 in a series of impressive stunts like pushing a wheelbarrow, carrying a man on his back and cooking an egg in mid-traverse. Reports in the British press paved the way for Blondin's arrival in London the following summer to recreate his feats in music halls before a huge painted backdrop of Niagara. He was joined in 1861 by his compatriot Jules Léotard, who swung and somersaulted high above the audience on a little-known apparatus called the trapeze.[46] The pair's popularity inspired British acrobats, contortionists and wire-walkers to don gorilla suits and adopt foreign-sounding names to perform updated versions of traditional 'Jocko' monkey sketches. These long-forgotten acts played on the dichotomies between humans and apes and verged in tone from farce to feats of strength. At the height of the craze, Londoners could watch performers like 'the celebrated Blondin and Léotard monkey', 'The gorilla monkey' and the cryptically advertised gorilla who would '"have his say" and "express his opinion" every night', a description suggesting a sketch in which the animal spoke directly to the audience.[47]

The most prolific performer to meld these overlapping sensations was Herr Shentini, an Englishman who assumed a foreign-sounding surname in the late 1840s to wire-walk in a hairy suit as 'the monkey man'.[48] He updated his act in August 1861 by dubbing himself 'the great gorilla, or man-monkey' and advertising his appearances with sober engravings of gorillas, probably torn from Du Chaillu's book. We have more than fleeting glimpses of Shentini thanks to the court proceedings that followed his alcohol-shortened engagement at Leicester in November 1864. These show that Shentini and his wife earned £2 15s per week to perform a sketch that played broadly on the theme of *Punch*'s Mr Gorilla. While Shentini's wages did not compare to those commanded by the biggest stars, he appeared regularly for many years in industrial centres in Scotland, Lancashire and Yorkshire and even occasionally in London, demonstrating that artistes could expect steady employment and decent, if not generous, wages from performing unsophisticated evolutionary acts.[49] The only visual representation I have found of these acrobatic gorillas appeared in the comic magazine *Fun* in 1886, a date that demonstrates the resilience of such acts (see Figure 2).[50]

Humour never directly addressed the tension and fear provoked by evolution. But to Charles Spurgeon, Britain's most renowned preacher, it showed an indecently blithe response to such ideas. Therefore he gave a heavily publicised lecture entitled 'The gorilla and the land he

2 Gorilla acrobats, *Fun* (10 March 1886).

inhabits' on 1 October 1861. Spurgeon had no academic credentials. He had begun preaching to London's largest Baptist congregation after an adolescent epiphany. As the crowds drawn by his oratory grew, he rented ever bigger venues until early 1861, when the Metropolitan Tabernacle opened in south London. The spare sobriety of the Greek portico on the building's facade disguised an interior that strongly resembled a lavish music hall, with filigreed ironwork, banked seats and balconies that accommodated over 5,000 worshippers. The tabernacle overflowed on the appointed evening. Though there is no hard evidence for the audience's motivations, we can postulate that many were seeking scriptural reassurance that humans and apes were unrelated. Du Chaillu, who sensed that Spurgeon's endorsement would be good publicity, sat on stage, while illustrations from his book were projected onto a screen and a stuffed gorilla stood in the pulpit with its arms outstretched in a sermonising pose. Spurgeon gave an impassioned speech that interwove preaching, political oratory and music hall performance. The audience cheered his anti-evolutionary bombast and curious insistence that Du Chaillu had opened West Africa for missionaries, though reports throughout Britain, the empire and the United States mocked his scientific ignorance.[51]

Spurgeon's Cnut-like attempts to hold back the tide of interest in gorillas opened him up to months of ridicule in *Fun*, beginning with a piece that reimagined his speech as a pun-filled music hall turn with jokes like 'now about M. Du Chaillu and his book; that is a volume of decided weight; I dropped it on my toes this morning, and so can speak from experience'.[52] The cartoons and jokes in *Fun* inspired similar ones in the United States, where many contemptuous reports were printed about Sturgeon's lecture.[53] He was equally ridiculed at London's Judge and Jury Music Hall, which staged a sketch whose cast included 'The female gorilla', 'Monsieur Shallyou D'Shallow' and 'Rev. Mr Scourgem'.[54] Pantomimes spanning the city from Whitechapel to the West End also featured comic allusions to gorillas, most notably Henry J. Byron's Drury Lane burlesque of *The Colleen Bawn*, Dion Boucicault's Irish murder melodrama. Among the up-to-date references was a song whose lyrics included the lines:

> I think I'll go to Londin [sic],
> Yes, take a foreign tour,
> See Leotard and Blondin,
> Likewise the Perfect Cure;
> Also the great Gorilla
> From foreign climates borne,
> That flash noble man-killer
> That now is so much worn.[55]

Byron's topical lyrics suggested that knowledge of gorillas and their acrobatic imitators had spread as far as the play's setting in rural western Ireland.

Fascination with gorillas was satirised by *Punch* at year's end with a cartoon entitled 'One good turn deserves another' that depicts Mr Gorilla as an itinerant organ grinder accompanied by a human on a chain who collects money from passers-by. The title was a pun on the word 'turn', as both a revolution of the organ's handle and a music hall performance.[56] A similar reference to folk culture appeared the following February, when Byron's 'A gorilla love song', coyly said to have been 'translated from the original gorillese', appeared in *Fun*. Its four-stanzas are filled with faux-folk idioms, recounting the seduction of an English lass by a gorilla. The animal begins by entreating her to 'list to my wooing / Hazel-haired girl' before shifting to more disturbing images of its desire to eat humans.[57] The pre-modern tone and breezy ballad-like structure nodded to the centuries-long fascination with freaks and monsters, while once again evoking the threatening notion that a woman had been raped by a gorilla. References to gorillas continued in comic magazines, pantomimes, popular plays, music halls and waxworks throughout the country for the rest of the decade.[58]

The gorilla was also an increasingly international character. In March 1862, *Fun* published a spoof which warned people travelling to Land's End, the Hebrides and Galway that they might encounter someone who was ignorant about the creature.[59] The joke vastly underestimated how far the sensation had spread. There was a much narrower vista for gorilla comedy in the United States, especially since the craze had coincided with the start of the Civil War. The charged climate meant that Mr Gorilla was initially reimagined as an angry, aggressive beast in cartoons suffused with long-standing anti-British sentiments and more contemporary anger about London's sympathy for the Confederacy. In November 1861, Mr Gorilla appeared in *Vanity Fair* as 'J. Africanus Gorilla A.M.', an educated ape that taught Confederate children about ignorance, hate and violence.[60] The magazine followed this in February 1862 with a cartoon entitled 'Gorilla Britannicus', presenting the animal as an aggressive and bellicose John Bull (see Figure 3).[61] Confederate newspapers responded by simianising President Abraham Lincoln's features, dubbing him the 'obscene ape of Illinois' and portraying him endorsing Barnum's 'What Is It?' as a candidate for the White House.[62] Anger predominated, but it was not the only tone in which Americans evoked gorillas. For instance, in September 1862 *Vanity Fair* published a cartoon depicting *Punch*'s gorilla sitting in a barber's chair as the nervous proprietor approaches. The subtitle shows that the animal has turned a banal quotidian exchange into a frightening confrontation. Mr Gorilla instructs the barber to 'Look out young man, how you treat my hair!', prompting the stammering reply 'Ye-ye-yes, Sir!', words so close to *Punch*'s cartoon that they

'Gorilla Britannicus', *Vanity Fair* (8 February 1862).

could have been imported directly.[63] Three months later, a New York magazine published a long spoof of an imagined address to the British Association, written in the gorilla's voice, in which the animal pleaded for acceptance.[64]

Mr Gorilla migrated to Australia more quickly and in much the same spirit as the British original. Australians had read about the Crystal Palace gorilla, bought Du Chaillu's book, seen *Punch*'s cartoons, read descriptions of Byron's stage turn, purchased the music to Paul's song, danced the 'Gorilla quadrille' and heard about Spurgeon's lecture within six weeks, essentially the time it took a ship to travel from Britain.[65] Australians adapted the character almost immediately for an original gorilla poem that was published in a Queensland newspaper in mid-June, and for jokes in *Melbourne Punch* the following month.[66] References to gorillas continued over the coming years in magazines and in stage plays like George Isaacs's 1865 *The Burlesque of Frankenstein; or the Man-Gorilla*, which included such lyrics as 'Adam was the first of men / gorilla's youngest brother', along with references to Darwin.[67] These evocations show that by the mid-1860s the gorilla was an international symbol through which to satirise evolutionary ideas.

None of the cartoonists, writers, artistes or acrobats who developed topical takes on the gorilla had ever seen a live animal. Relatively few monkeys and apes survived extended sea voyages or lived long in Britain's climate. Still, circus and menagerie owners passed off various large apes as the fabled beast in order to sate this desire. The most prominent of them was G.A. Van Hare, who was managing London's Surrey Gardens Theatre in the late 1850s; there he hired a French high-wire walker, dubbed her the 'Female Blondin' and had her perform on a wire stretched across the Thames in a spoof of her namesake's far riskier traverses of Niagara. He then trained a baboon to walk on a high-wire in front of lantern slide images of the real Blondin, much as costumed human acrobats were doing throughout the country.[68] Van Hare claimed that soon thereafter he visited Spain, where an American sailor told him 'to get one of them gorillas there has been so much talk about in the States ... Some Frenchman, I think, brought a stuffed animal almost like a man, but much larger'.[69] Van Hare's ignorance about the animal is as hard to swallow as the details of his subsequent West African gorilla hunt. Humbug aside, it is certain that at the start of 1862 he was exhibiting a pair of trained apes and introduced 'Hasan, the Gorilla Chief', his most important act, in March of the following year.[70]

It is virtually certain that Hasan was a large ape, but not a gorilla. Van Hare's story demonstrates that audiences were so hungry for gorillas that they could be satisfied with any large beast, as long as it had been trained to give a topical performance. At the same time, the relationship between showmen and audiences was not a simple interplay of trickery and gullibility, since both parties knew that entertaining topicality always trumped scientific verisimilitude. As a result, many supposed gorillas travelled with British circuses over the coming years including 'the Great Pongo, or Ethiopian Savage', the unfortunately named 'Dingy', an 'equestrian gorilla' that turned somersaults on the back of a moving horse, and one whose skeleton, in the Norwich Castle Museum, has since been proved to be that of an actual gorilla.[71]

It is just as telling to see where gorillas made relatively little impact. Despite the animal's inroads into music hall, burlesque and pantomime, it was almost absent from the minstrel shows that were popular in 1860s Britain. This is initially surprising, given the racist associations that were commonly made, implicitly or explicitly, between the animal and black people. Minstrel shows had originated in the United States, where images of contented plantation life were intended in part to prop up the slave system. The gorilla craze was reflected in a version of the song 'The whole hog or none' by the Great Mackney, Britain's most famous minstrel. He had updated the lyrics with references to Léotard and lectures about apes.[72] Equally telling is a comic exchange written by Charles Dickens in which two men observing seaside minstrels ingeniously subvert the relationship between the races. One questions why artistes don blackface to perform quotidian music hall songs, before pleading the cause of white performers by asking 'Am I not a man and a brother?'[73] Such examples aside, the fairly rigid structure of minstrel shows, the fixed roles allotted to performers and the humour that turned on the discord between the 'backward' slave characters and their sophisticated patter and musical abilities left little room for gorillas. Press reports suggest that when gorilla performers appeared with British minstrel shows, like the contortionist who was described as having 'an unmistakable resemblance' to Du Chaillu's animal in 1863, they were an added attraction, rather than being fully incorporated into the troupe. Nonetheless, such acts must still have provided fodder for racist associations between blacks and gorillas.[74]

The relative absence of gorillas in British minstrelsy also reflects the country's ethnic make-up and historical prejudices. Minstrel shows mirrored American society and bore little resemblance to the day-to-day experience of life in mid-Victorian Britain. There were comparatively few black people in the British Isles, though many were encompassed within the emerging empire and these people had often been exhibited for supposedly educational purposes. The Irish were Britain's semi-human underclass.

Their Roman Catholic faith signalled ignorance, while they were perceived as dissolute, ill-educated and potentially treasonous. They were constantly depicted in cartoons and caricatures with explicitly simian features, and in October 1862 *Punch* even declared them the missing link between gorillas and humans. The magazine justified the epithet bitingly, by explaining how they inhabited the slums of Liverpool and London – areas as dark and threatening to Victorians as West Africa – and described them as 'a climbing animal, [that] may sometimes be seen ascending a ladder laden with a hod of bricks', in reference to the navvies employed in the building trades throughout England.[75]

The publisher Thomas Nelson reacted to the gorilla sensation by offering the prodigious Scottish author R.M. Ballantyne £80 to write a novel for young people based on Du Chaillu's book, so long as he completed it in time for the profitable Christmas sales. *The Gorilla Hunters* was a follow-up adventure for the young trio whom Ballantyne had introduced in his 1858 novel *The Coral Island*, and marks the animal's literary debut. The friends hunt vainly for a gorilla in Africa, causing the narrator Ralph Rover and his dispirited companion Peterkin Gay to become vexed that showmen have tricked them into searching for a non-existent creature:

> 'I don't believe there's such a beast as a gorilla at all; now, that's a fact … Ralph, it is my belief, I tell you, that the gorilla is a regular sell – a great, big, unnatural hairy *do*!'
> 'But I saw the skeleton of one in London.'
> 'I don't care for that. You may have been deceived, humbugged.'[76]

Soon thereafter, the trio approach a dense forest, causing Peterkin to state excitedly, 'it is probable that we may find Mister Gorilla there', an allusion to the sensation's most evocative image that demonstrates how the term had entered the argot and imaginations of the schoolboys and armchair adventurers who read Ballantyne's novels.[77]

A more formidable incredulity was expressed by Charles Kingsley in his moral fable *The Water Babies*, which was serialised in *Macmillan's Magazine* from August 1862 to the following March. Kingsley had been a confidante of Darwin, but the novel showed his increasing difficulties in reconciling faith with evolution.[78] The following passage, one of many that could be included, evokes the nose-thumbing attitude towards scientific knowledge that was apparent in Kingsley's Dundreary spoof: 'you do not know what Nature is, or what she can do; and nobody knows; not even Sir Roderick Murchison, or Professor Owen, or Professor Sedgwick, or Professor Huxley, or Mr. Darwin, or Professor Faraday, or Mr. Grove, or any other of the great men whom good boys are taught to respect'.[79] The passage suggested that scientists' intellectual pride had verged into hubris.

Ballantyne and Kingsley are not widely read today. The most lasting literary evocation of gorillas in print comes from Huxley's 1863 tome *Evidence as to Man's Place in Nature*, a compendium of his lectures for working men. While Huxley was fierce and uncompromising with academic peers, he talked to artisans in commonplace language and used showman's stage tricks like draping his arms over an ape's skeleton, a gesture of casual intimacy that owed more to the music hall than to the laboratory.[80] In the book, Huxley's tone veers from apologia to assault by way of mocking incredulity, while his greatest bile is expended on Du Chaillu:

> not because I discern any inherent improbability in his assertions respecting the man-like Apes; nor from any wish to throw suspicion on his veracity; but because, in my opinion, so long as his narrative remains in its present state of unexplained and apparently unexplainable confusion, it has no claim to original authority respecting any subject whatsoever. It may be truth, but it is not evidence.[81]

The final sentence was an often-repeated wholesale dismissal of the explorer, though it is the illustration in the book's frontispiece that has become one of the most widely recognised scientific images of all time. Benjamin Waterhouse Hawkins drew the skeletons of a gibbon, an orang-utan, a chimpanzee, a gorilla and a human walking in profile across the page, implying a direct evolutionary line from apes to modern humans. The image has been reprinted exactly and in endlessly fanciful ways for 150 years.

A second comment on the gorilla debate that still resonates was voiced the following November by the future Prime Minister Benjamin Disraeli, even if, as his biographer stated, 'the moral and intellectual problems which vexed the graver portion of the Victorian governing class were of no interest' to him.[82] In other words, he was like a large segment of the population who were aware of the debate's general outlines without being much invested in it. At a meeting of the Society for Increasing the Endowments of Small Livings in the Diocese of Oxford – a more recondite Anglican body is hard to imagine – Disraeli declared that 'the question between science and religion is whether man is an ape or an angel? I, am on the side of the angels.'[83] What seems like a categorical dismissal, proclaimed appositely in Bishop Wilberforce's see, actually cloaked shrewd political nous by affirming the traditional alignment between the Church of England and the Conservative Party.

It is also a fitting phrase with which to close this chapter, because it shows that the acrimonious scientific battle had created a widely diffused, if somewhat simplistic, knowledge of evolutionary ideas. Evolutionary debates and diatribes, along with gentler quips and satires, had invaded the respectable middle class through the British Association for the Advancement of Science, *Punch*, music hall and other means. Humorous responses – some scientific and learned, others absurd and grotesque – had conflated and confused Darwinian theories with potentially explosive ideas about kinship with African gorillas. These had quickly transformed gorillas from dumb animals into sentient proto-humans who spoke directly to their audiences. It was a nightmare vision to those whose opinions or faith were inflexible, but a farcical one to the larger and less doctrinaire body of people who only wanted to be entertained. Gorillas became very familiar in British popular culture in the 1860s, being portrayed equivocally either as degrading visions of humanity or through humour that played on the racist unease they had engendered. It was an evasive point of view that acknowledged the fierce debates, while permitting audiences to laugh at them indirectly. Laughter mitigated the threat posed by gorillas. It is the predominant tone in which prehistory has been represented in popular culture ever since.

Notes

1 'A new race of monkeys – the gorillas', *Harper's Weekly* (5 March 1859), p. 148.
2 'The "gorilla" at the Crystal Palace', *Morning Chronicle* (8 November 1858), p. 5; 'The gorilla', *Children's Friend* (1 December 1859), p. 273.
3 'The Alhambra', *Era* (13 March 1859), p. 11; 'Highbury Barn', *Era* (6 March 1859), p. 8.
4 Charles Darwin, *On the Origin of Species* (London: John Murray, 1859).
5 Janet Browne, *Darwin's 'Origin of Species': A Biography* (Toronto: Douglas & McIntyre, 2006), pp. ix–x.
6 Alvar Ellegård, *Darwin and the General Reader* (Chicago: University of Chicago Press, 1990), p. 24; Adrian Desmond and James Moore, *Darwin's Sacred Cause: Race, Slavery and the Quest for Human Origins* (London: Allen Lane, 2009), p. 328.
7 Ellegård, *Darwin*, pp. 24–7; and Edward Caudill, 'The bishop-eaters: the

publicity campaign for Darwin and *On the Origin of Species*', *Journal of the History of Ideas*, 55:3 (1994), pp. 441–60.
8 Nicolaas A. Rupke, *Richard Owen: Victorian Naturalist* (New Haven: Yale University Press, 1994), p. 298.
9 Ian Hesketh, *Of Apes and Ancestors: Evolution, Christianity and the Oxford Debate* (Toronto: University of Toronto Press, 2009), pp. 76–7; Desmond and Moore, *Darwin's Sacred Cause*, p. 346; Paul White, *Thomas Huxley: Making the 'Man of Science'* (Cambridge: Cambridge University Press, 2003), pp. 51–7; and 'Samuel Wilberforce', in Sidney Lee (ed.), *Dictionary of National Biography*, vol. XXI (London: Smith Elder, 1909), p. 207.
10 The quotation is from Jonathan Conlin, *Evolution and the Victorians* (London: Bloomsbury, 2014), pp. 95–6. See also Hesketh, *Of Apes*, pp. 82–95; Harpocrates, 'The Darwin "developments" at Oxford', *Morning Chronicle* (9 July 1860), p. 7; and 'Literary', *New York Tribune* (4 August 1860), p. 6.
11 Adrian Desmond, *Huxley: From Devil's Disciple to Evolution's High Priest* (Reading: Addison Wesley, 1997), pp. 208–11 and 292–3; 'Professor Huxley on the Negro', *John O-Groat Journal* (13 June 1867), p. 4.
12 'Unnatural selection and improvement of species', *Punch* (10 November 1860), p. 182.
13 *Charles Kingsley: His Letters and Memories of his Life, Edited by his Wife* (New York: Scribner Armstrong, 1877), pp. 322–5. Reports about the speech included 'Lord Dundreary's opinions on the "Darwinian Theory"', *Leeds Mercury* (9 October 1862), p. 4. See also Rupke, *Owen*, p. 298.
14 Pycroft's satire is in Frederick Burkhardt (ed.), *The Correspondence of Charles Darwin*, vol. XI (Cambridge: Cambridge University Press, 1999), pp. 769–75. See also 'George Pycroft', *The Times* (4 April 1894), p. 10.
15 Ronald L. Numbers, *Darwinism Comes to America* (Cambridge, Massachusetts: Harvard University Press, 1998), pp. 1–2 and 30–75; 'Introduction' in A. Hunter Dupree (ed.), *Darwiniana: Essays and Reviews Pertaining to Darwinism by Asa Gray* (Cambridge, Massachusetts: Harvard University Press, 1963), pp. ix–xxiii.
16 'A new African explorer', *New York Tribune* (27 August 1859), p. 4; 'American Geographical and Statistical Society', *New York Tribune* (6 January 1860), p. 8.
17 'American theatricals', *Era* (1 April 1860), p. 5.
18 James W. Cook, *The Arts of Deception: Playing with Fraud in the Age of Barnum* (Cambridge, Massachusetts: Harvard University Press, 2001), pp. 132–6 and 151–3; Jane R. Goodall, *Performance and Evolution in the Age of Darwin: Out of the Natural Order* (London: Routledge, 2002), pp. 53–7; Monte Reel, *Between Man and Beast: An Unlikely Explorer, the Evolution Debates, and the African Adventure that Took the Victorian World by Storm* (New York: Doubleday, 2013), pp. 97–109.
19 'The gorilla at home', *Leisure Hour* (18 April 1861), pp. 253–5.
20 'Gorilla Du Chaillu', *Harper's Weekly* (6 April 1861), p. 211.
21 'Royal Institution', *Morning Chronicle* (19 March 1861), p. 6.

22 Gowan Dawson, *Darwin, Literature and Victorian Respectability* (Cambridge: Cambridge University Press, 2007), pp. 60–9.

23 The quotation is from Paul Du Chaillu, 'To the editor', *The Times* (5 July 1861), p. 6. See also Richard Burton, 'To the editor', *The Times* (8 July 1861), p. 10; Stuart McCook, '"It may be truth, but it is not evidence": Paul Du Chaillu and the legitimation of evidence in the field sciences', *Osiris*, 11 (1996), pp. 185–95; 'M. Du Chaillu's collection of gorillas at the Royal Geographical Society', *Era* (5 May 1861), p. 15; 'Ethnological Society', *Morning Chronicle* (4 July 1861), p. 2; 'M. Chaillu and his gorillas', *Era* (14 July 1861), p. 10; untitled, *John Bull* (20 July 1861), p. 464; 'The gorilla in the British Museum', *Bell's Life in London* (28 July 1861), p. 8; and 'The British Association', *Liverpool Mercury* (2 September 1861), p. 2.

24 'Monkeyana', *Punch* (18 May 1861), p. 206; Rupke, *Owen*, pp. 298–9.

25 'The lion of the season', *Punch* (25 May 1861), p. 213; and 'M. Du Chaillu's African adventures', *Literary Gazette* (18 May 1861), p. 459.

26 Peter Bailey, 'Conspiracies of meaning: music-hall and the knowingness of popular culture', *Past & Present*, 144 (1994), pp. 138–70.

27 'Epsom races', *Era* (2 June 1861), p. 13.

28 'The gorilla', *West Middlesex Advertiser and Family Journal* (10 August 1861), p. 3.

29 'Midsomer Norton', *Bath Chronicle* (8 August 1861), p. 5.

30 'Have you seen the gorilla?', *Shoreditch Observer* (9 November 1861), p. 2. The cases are discussed in Heinz K. Henisch and Bridget A. Henisch, *Positive Pleasures: Early Photography and Humor* (University Park: Pennsylvania State University Press, 1998), pp. 14–16.

31 Untitled, *Fun* (9 November 1861), p. 82.

32 'Law and police', *Era* (6 April 1862), p. 7; 'The week in Parliament', *Norfolk News* (28 February 1863), p. 2; 'Miners' demonstration at Normanby', *Northern Echo* (21 July 1873), p. 3.

33 J.R. Planché, *The Recollections and Reflections of J.R. Planché*, vol. II (London: Tinsley Brothers, 1872), pp. 217–19; Jeffrey Richards, *The Golden Age of Pantomime: Slapstick, Spectacle and Subversion in Victorian England* (London: I.B. Taurus, 2015), pp. 65–123.

34 'Henry James Byron', in Leslie Stephen and Sidney Lee (eds.), *Dictionary of National Biography*, vol. III (London: Smith Elder, 1908), pp. 607–9.

35 'Lyceum Theatre', *Morning Chronicle* (20 June 1861), p. 3.

36 The quotation is from 'Adelphi', *Players* (1860–61), p. 14. See also 'Adelphi New Theatre Royal', *Era* (30 June 1861), p. 8; and Henry R. Addison, *The Blue-Faced Baboon* (London: John Dicks, 1884?).

37 'Adelphi Theatre', *London Standard* (2 July 1861), p. 6. See also J.L. Toole, *Reminiscences of J.L. Toole Related by Himself and Chronicled by Joseph Hatton* (London: George Routledge and Sons, 1892), p. 126.

38 'Adelphi', *Era* (7 July 1861), p. 10; 'Dublin – Queen's Theatre', *Era* (22 September 1861), p. 11; 'Theatre Royal', *Belfast Mercury* (4 October

1861), p. 2; 'Edinburgh – Queen's Theatre', *Era* (13 October 1861), p. 11.
39 Henry J. Byron and F. Musgrave, 'Mr Gorilla, or the lion of the season' (London: Hopwood and Crew, 1861), British Library, Music Collections Hirsch M.1317(9).
40 'Christmas amusements', *Birmingham Daily Post* (7 December 1861), p. 4.
41 See for instance 'Mr and Mrs Howard Paul', *Morning Chronicle* (3 January 1862), p. 6; 'Mr and Mrs Howard Paul', *Era* (5 January 1862), p. 10; 'For one night only', *Burnley Advertiser* (11 January 1862), p. 2; 'Saturday evening concerts', *Liverpool Mercury* (13 January 1862), p. 1; 'Rotundo, Dublin', *Freeman's Journal and Daily Commercial Advertiser* (13 January 1862), p. 1; 'Mr and Mrs Howard Paul's entertainment', *Cork Examiner* (31 January 1862), p. 3; and 'Exchange Rooms', *Dundee Courier* (8 March 1862), p. 1.
42 The quotation is from C.H.R. Marriott, 'The Gorilla quadrille', British Library, Music Collections h.721.o.(5). See also 'Highbury Barn', *London Evening Standard* (16 November 1861), p. 1; 'Dramatic performance at the Wolverhampton Theatre', *Era* (22 February 1863), p. 13; 'Redhill', *Sussex Advertiser* (12 December 1865), p. 3. See also 'Arms and the man', *Punch* (21 September 1861), p. 115; 'The November magazines', *Bradford Observer* (7 November 1861), p. 7; and 'The gorilla's dilemma', *Punch* (18 October 1862), p. 164.
43 'The gorilla's dilemma', *Littell's Living Age* (29 November 1862), p. 43.
44 'M. Du Chaillu and the gorilla', *Manchester Courier and Lancashire General Advertiser* (7 September 1861), p. 7. See also 'The gorilla country in Africa', *Morning Chronicle* (2 May 1861), p. 6; 'Mr Du Chaillu's *Explorations*', *The Times* (8 May 1861), p. 13; 'M. Du Chaillu's lecture on the gorilla', *Glasgow Herald* (10 October 1861), p. 2; and 'M. Du Chaillu on the gorilla', *Leeds Mercury* (6 November 1861), p. 3.
45 See for instance 'Mr Hawkins's lectures', *Morning Post* (17 May 1861), p. 6; 'Mr Waterhouse Hawkins on the gorilla', *Ipswich Journal* (26 October 1861), p. 6; and 'Lecture on the gorilla', *Aris's Birmingham Gazette* (9 November 1861), p. 7.
46 See for instance 'The Niagara rope walker', *Era* (7 August 1859), p. 10; 'Young Blondin', *Era* (1 July 1860), p. 1; 'Léotard, the Flying Man', *Era* (19 May 1861), p. 10; and 'Léotard at the Alhambra', *Era* (26 May 1861), p. 7.
47 The quotations are respectively from 'South London Music Hall', *Era* (1 September 1861), p. 10; 'Victoria Theatre', *Era* (27 October 1861), p. 8; and 'The Bedford Music Hall', *Era* (17 November 1861), p. 16.
48 'Burnley', *Era* (18 March 1849), p. 12.
49 'A "gorilla" in the witness box', *Chester Chronicle* (24 December 1864), p. 6; 'William Shentini v. William Paul', *Leicester Journal* (16 December 1864), p. 6. See also 'Lancaster', *Era* (20 November 1859), p. 12; 'Theatre Royal, *Burnley Advertiser* (5 October 1861), p. 1; 'Northampton, Shakespeare Saloon', *Era* (19 January 1862), p. 12; 'Rotherham – Victoria Music Hall',

Era (8 February 1863), p. 13; 'New Adelphi Theatre', *Liverpool Mercury* (26 October 1863), p. 6; 'The Alhambra, Shoreditch', *Era* (30 October 1864), p. 11; 'Grimsby – Walker's New Music Hall', *Era* (26 March 1865), p. 14; 'Halifax – Circus Ward's End', *Era* (11 February 1866), p. 13; 'Sheffield – Canterbury Hall', *Era* (25 February 1866), p. 14; and 'The York Theatre', *Yorkshire Gazette* (27 August 1867), p. 9.

50 'The acrobatic ape', *Fun* (10 March 1886), p. 108.

51 See for instance Reel, *Man and Beast*, pp. 164–7 and 191–8; 'Mr Spurgeon on the gorilla', *London Daily News* (2 October 1861), p. 6; 'Mr Spurgeon on the gorilla', *Morning Post* (2 October 1861), p. 6; 'Mr Spurgeon and the gorilla', *Belfast Mercury* (11 October 1861), p. 3; 'London gossip', *Limerick Reporter and Tipperary Vindicator* (11 October 1861), unpaginated; 'Amusement', *Cleveland Morning Leader* (Ohio) (23 October 1861), unpaginated; and 'Domestic affairs', *Argus* (Melbourne) (16 December 1861), p. 6.

52 The quotation is from 'A pretty go-realla', *Fun* (12 October 1861), p. 34. See also 'Touchstone's telegrams', *Fun* (19 October 1861), pp. 46 and 50; 'The Lord Mayor's show', *Fun* (9 November 1861), p. 74; 'Moral', *Fun* (16 November 1861), p. 84; 'Literary intelligence', *Fun* (7 December 1861), p. 116; 'Spurgeon at home', *Fun* (25 January 1862), p. 190; and 'The Crystal Palace', *Fun* (8 February 1862), p. 204.

53 'Mr Spurgeon on the Gorilla!', *Ballou's Dollar Monthly Magazine* (July 1862), p. 99.

54 'The female gorilla', *Era* (13 October 1861), p. 1.

55 The quotation is from Henry J. Byron, *Miss Eily O'Connor, A New and Original Burlesque Founded on the Great Sensation Drama of the Colleen Bawn* (London: Thomas Hailes Lacy, 1861?), pp. 24–5. See also 'Drury Lane', *Athenaeum* (7 December 1861), p. 773; and 'Royal English Opera', 'Drury Lane Theatre Royal', 'Eastern Opera House Pavilion', 'Standard Theatre, Shoreditch' and 'Grecian Theatre', all *Era* (22 December 1861), p. 8.

56 'One good turn deserves another', *Punch's Almanack for 1862* (1861), unpaginated. *Punch's Almanack*s were published in December, in time for Christmas sales. So the *Almanack* for 1862 actually appeared in the last weeks of 1861.

57 'A gorilla love song', *Fun* (22 February 1862), p. 231.

58 See for instance 'Alhambra', *Era* (2 March 1862), p. 8; 'Birmingham – Holder's Concert Hall', *Era* (4 January 1863), p. 11; Frank Burnand, 'Monkeanna: or the white witness', *Punch* (21 February 1863), pp. 71–2; 'Mr W. Patterson', *Era* (8 November 1863), p. 1; 'Middlesbrough – Theatre Royal', *Era* (3 January 1864), p. 13; 'James Chester', *Era* (19 August 1866), p. 1; 'Holder's Concert Hall', *Era* (9 December 1866), p. 11; 'Tunstall – Prince of Wales Theatre', *Era* (22 December 1867), p. 14; 'Merthyr Tydfil', *Era* (28 June 1868), p. 13; 'Wolverhampton – Theatre Royal', *Era* (22 November 1868), p. 15; 'Sinclair's New Music Hall, Carlisle', *Era* (11 July 1869), p. 8; and 'Wakefield – Corn Exchange', *Era* (31 January 1869), p. 13.

59 'To persons about to travel', *Fun* (15 March 1862), p. 253.

60 'J. Africanus Gorilla A.M.', *Vanity Fair* (2 November 1861), p. 207. See also Jane E. Brown and Richard Samuel West, 'William Newman (1817–1870): a Victorian cartoonist in London and New York', *American Periodicals*, 17:2 (2007), pp. 167–73; 'Letter to King James of America', *Vanity Fair* (21 January 1860), p. 59; 'The gorilla war', *Harper's Weekly* (17 August 1861), p. 514; and Robert J. Scholnick, 'The fate of humor in a time of civil and cold war: *Vanity Fair* and race', *Studies in American Humor*, 3:10 (2003), pp. 21–42.

61 Stephen Bobbett-Hooper, 'Gorilla Britannicus', *Vanity Fair* (8 February 1862), p. 71.

62 The quotation is from 'To Union men', *Harper's Weekly* (27 August 1864), p. 547. See also 'South-side despatch', *Head Quarters* (16 October 1861), p. 4. For the 'What Is It?' see James K. Lively, 'Propaganda techniques of Civil War cartoonists', *Public Opinion Quarterly*, 6:1 (1942), pp. 99–106. For Lincoln's simian features see Christopher Kent, 'War cartooned/cartoon war: Matt Morgan and the American Civil War in *Fun* and *Frank Leslie's Illustrated Newspaper*', *Victorian Periodicals Review*, 36:2 (2003), p. 160.

63 Untitled, *Vanity Fair* (13 September 1862), p. 123. The artist may also have known 'Cruel', *Fun* (16 November 1861), p. 92.

64 'The gorilla's petition', *Albion* (27 December 1862), p. 613.

65 See for instance 'Discoveries in Central Africa', *Argus* (Melbourne) (7 May 1861), p. 6; 'The last lion', *Cornwall Chronicle* (Tasmania) (7 August 1861), p. 4; 'Arrival of the Salsette', *Empire* (Sydney) (15 August 1861), p. 2; untitled, *Inquirer and Commercial News* (Perth) (25 September 1861), p. 2; 'Western Australia', *Courier* (Brisbane) (26 October 1861), p. 4; 'Our London Correspondence', *Empire* (Sydney) (19 December 1861), p. 2; 'Sydney Music Hall', *Sydney Morning Herald* (19 February 1862), p. 12; and 'New books and music', *Sydney Morning Herald* (19 July 1864), p. 6.

66 'A true story', *Maryborough Chronicle* (Queensland) (27 June 1861), unpaginated; 'The gorilla', *Melbourne Punch* (18 July 1861), p. 151; 'An alarming visitor', *Melbourne Punch* (25 July 1861), p. 160.

67 The quotation is from George Isaacs, *Rhyme and Prose: A Burlesque and its History* (Melbourne: Clarson, Shallard, 1865), p. 102. See also for instance 'Political dialogues', *Courier* (Brisbane) (1 March 1864), p, 2; 'Quadrupled ingenuity', *Sydney Punch* (11 June 1864), p. 20; 'Foley's', *Sydney Morning Herald* (10 July 1866), p. 8.

68 'The "female Blondin" at Cremorne', *Era* (18 August 1861), p. 10.

69 W.G. Van Hare, *Fifty Years of a Showman's Life* (London: W.H. Allen, 1888), pp. 17–70; the quotation is on pp. 143–4. See also 'Van Hare's visit to Morocco', *Era* (18 August 1861), p. 1.

70 'Van Hare's lot of startling novelties', *Era* (19 January 1862), p. 1; 'A challenge to all nations for 1,000 guineas', *Era* (8 March 1863), p. 15; 'Hasan the gorilla chief', *Era* (8 March 1868), p. 15; Van Hare, *Fifty Years*, pp. 282–98.

71 See for instance 'Manders' Excelsior Menagerie', *Norfolk News* (1 March 1862), p. 4; 'Dingy the young male gorilla', *Era* (31 May 1863), p. 1;

'Barnsley – Queen's Theatre', *Era* (7 February 1864), p. 11; 'Middlesbrough – Royal Albert Theatre', *Era* (31 March 1867), p. 13; 'Leeds – Newsome's Circus', *Era* (7 April 1867), p. 12; and 'Agricultural Hall', *Era* (7 February 1869), p. 6.

72 J.S. Bratton, 'English Ethiopians: British audiences and black-face acts', *Yearbook of English Studies*, 11 (1981), p. 142.

73 Philip Collins, 'Some unpublished comic duologues of Dickens', *Nineteenth-Century Fiction*, 31:4 (1977), p. 443.

74 The quotation is from 'Bolton – Theatre Royal', *Era* (8 July 1863), p. 11. See also 'Accrington', *Era* (24 May 1863), p. 11; 'Dudley – the Music Hall', *Era* (25 December 1864), p. 12; 'Mr F. Simmonds, the original gorilla', *Era* (8 January 1865), p. 1; 'Leamington – Spa Promenade', *Era* (12 February 1865), p. 13; 'Grantham – Christy's Minstrels', *Era* (19 March 1865), p. 13; 'Wolverhampton – Prince of Wales Concert Hall', *Era* (2 April 1865), p. 14; 'Brighton – Canterbury Hall', *Era* (28 January 1866), p. 12; and 'Longton – People's Music Hall', *Era* (21 June 1868), p. 13.

75 The quotation is from 'The missing link', *Punch* (18 October 1862), p. 165. See also L.P. Curtis, *Apes and Angels: The Irishman in Victorian Caricature* (Washington, DC: Smithsonian Press, 1997).

76 R.M. Ballantyne, *The Gorilla Hunters* (London: T. Nelson and Sons, 1861), p. 9.

77 Ballantyne, *Gorilla*, p. 120. See also Eric Quayle, *Ballantyne the Brave: A Victorian Writer and his Family* (London: Hart-Davis, 1967), p. 146; 'Christmas books', *Examiner* (21 December 1861), p. 6; and 'Literary notices', *Caledonian Mercury* (23 December 1861), p. 3.

78 Amanda Hodgson, 'Defining the species: apes, savages and humans in scientific and literary writing of the 1860s', *Journal of Victorian Culture*, 4:2 (1999), pp. 237–41.

79 Charles Kingsley, *The Water Babies* (London: Macmillan, 1910), pp. 69–70.

80 Conlin, *Evolution*, p. 98; Desmond, *Huxley*, pp. 639–40; Leonard Huxley (ed.), *The Life and Letters of Thomas Henry Huxley* (Cambridge: Cambridge University Press, 2012), pp. 380–2.

81 Thomas Henry Huxley, *Evidence as to Man's Place in Nature* (New York: Appleton, 1863), p. 68.

82 Robert Blake, *Disraeli* (London: Prion, 1998), p. 505.

83 Browne, *Darwin's 'Origin'*, p. 99; 'Dressing for an Oxford bal masqué', *Punch* (10 December 1864), p. 239.

3

The parents of Adam and Eve: missing links

In late 1870, Thomas Henry Huxley published a reflection about his decade-old laudatory review of *On the Origin of Species*, apologising to readers that 'if you find its [the review's] phraseology, in some places, to be more vigorous than seems needful, recollect that it was written in the heat of our first battles', referring to the dispute over Darwinian theory.[1] As Huxley indicated, the animus of the early 1860s had given way to a general acknowledgement of evolution, even if scientists still debated the theory's finer points. Darwin, the lightning rod in earlier debates, had spent the intervening years in affluent rural seclusion. Then in February 1871, he published *The Descent of Man, and Selection in Relation to Sex*, in which he argued that humans had evolved from primitive species, placing sexual desire at the centre of evolution for the first time. Passions were rekindled, and the book was received in 'a storm of mingled wrath, wonder and admiration', though many scientific responses were too arcane for the general public.[2] Darwin followed *Descent* in 1872 with *Expression of the Emotions in Man and Animals*, in which he explored the origins of many human gestures and expressions.[3]

Once again, humour helped mitigate Darwin's unsettling ideas. He was caricatured repeatedly in comic magazines as a lustful simian, while Cambridge undergraduates dangled a stuffed ape over the Senate House's balcony and chanted 'monkey, monkey' as he received an honorary degree in 1875.[4] It was a typically irreverent student display created by young men who had grown up amid evolutionary debates and satires. A more sedate expression of the wider accord between science and faith was evident in the reaction to a paper that the Assyriologist George Smith read in December 1872 to an audience that included the Archbishop of Canterbury and the Prime Minister William Ewart Gladstone, a dedicated Classical scholar and devout Anglican. Smith announced that he had deciphered cuneiform inscriptions from present-day Iraq to reveal the *Epic of Gilgamesh*, a previously unknown account of the Flood that predated the Bible by twenty centuries. Gladstone reacted by telling the room, 'we shall be permitted to know a great deal more than our forefathers in respect of the early history of mankind', while the public subscribed to a fund that sent Smith back to search for more of the story.[5]

Smith's evidence was cast in hard mud brick: tangible proof that still evaded evolution's champions. No intermediary species, the so-called 'missing link', between apes and humans had been found. Geologists had coined the term in the 1850s to denote gaps in the sedimentary record and it had subsequently been co-opted by evolutionists. While humanity's most ancient past remained elusive, evidence of European prehistory was increasing. The polymathic scientist Sir John Lubbock first used the term 'cave man' in 1865 in a popular exploration of prehistory that remained in print for fifty years. Lubbock employed the term to describe the ancient Europeans whose remains were being uncovered in ever greater numbers. Prehistoric collections were displayed at the 1867 Paris Exhibition, Scandinavian archaeologists developed a chronological structure for the European Iron Age, and many popular and heavily illustrated works on prehistory were published.[6] Such attempts to promote scientifically accurate knowledge of the ancient past continued to coexist with light-hearted invocations of evolution, like the poetic attack on a Sheffield city councillor in November 1872:

> To the man from the monkey (so Darwin believes)
> This was the original track;
> And now Mr Councillor Hutchinson grieves,
> To the monkey he cannot go back.
> With a tail to be twisting while munching his bun
> What monkey could furnish more excellent fun?[7]

This brief paragraph is intended to show that, despite gaps in knowledge of the very distant past – lacunae that largely remain to this day – much was understood by scientists, popularised in books and magazines and casually invoked by the public.

Scientific knowledge was popularly distilled into a simplistic conceit that there had been a single missing link, while the lack of concrete evidence created an imaginary space that could be filled with all manner of humbugs and humour.[8] Missing-link jokes abounded in comic magazines like *Judy*, which speculated in October 1870 that 'Doctor Darwin has at length discovered the Missing Link at an Eastern music hall', referring to popular fears about the lawless, wild state of London's slums.[9] The joke was echoed in 1872, when *Punch* challenged the Welsh-American explorer Henry Morton Stanley, who had famously located Dr Livingstone twelve months earlier, to find 'the missing link in the last London fog'.[10] Most influentially, the fascination with missing links inspired Howard Paul to update the song and sketch based on *Punch*'s 'Mr Gorilla' cartoon that he had performed in the 1860s. In late 1872 he premiered 'The parents of Adam and Eve: according to Darwin', a song whose title explicitly challenged the Biblical narrative and invoked interbreeding between apes and humans in a way that would have been incendiary a decade earlier. The lyrics were a long commentary on *Descent*, whose basic argument Paul expected his audiences to understand, beginning:

> A book has been written renowned
> For strangest of things to believe
> The grave Dr Darwin has found
> The parents of Adam and Eve.
> Old Adam it seems had a father,
> And Eve a mother had she,
> They cracked cocoanuts and they rather preferred to live up in a tree.[11]

Paul had performed his earlier sketch clad as a Mr Gorilla, when it had been the most potent evolutionary satire. He now took the stage as Lord Dundreary, the brainless aristocrat from popular theatre. He was joined in mid-song by a performer clad in a gorilla suit and a ball dress, with whom he danced a polka in a sublimated version of the inter-species mating at the heart of *Descent*. This inspired a Devon newspaper to pun that 'When Mr Howard Paul is on the stage with his ancestor, the ape, what well-known book do they represent? Answer – Hims Ancient and Modern – with an Appendix'.[12] Paul performed the song throughout Britain until at least 1874 and perhaps even in the United States, where the sheet music was published the following year.[13]

Paul's Dundreary costume was doubly punning, since it also spoofed the 1872 Covent Garden pantomime *Babil and Bijou*, the most expensive stage production ever mounted in Britain and a colossal flop. The plot, which was partly devised by James Planché, who had written a pair of gorilla burlesques in 1861, concerned a deposed fairy queen's quest to regain her throne. It included a scene entitled 'The nine ages of man', in which the stage was filled with a pyramidal representation of human development – whose title and visuals spoofed Shakespeare and Huxley in equal measure – with an ape at the bottom and Dundreary at the pinnacle.[14] Other reflections of the ongoing fascination with gorillas were heard in London plays and pantomimes, missing-link songs and revivals of 'The gorilla quadrille', the dance conducted by a band leader clad as Mr Gorilla, while acrobats continued exploiting the character's physical traits, and circuses and menageries vaunted their living animals.[15]

Darwinian ideas also gained credence in the American imagination in the decades after the Civil War, which ended in 1865, though not primarily, as might have been anticipated, to support racist beliefs. Legislative attempts to ban the teaching of evolution, and to prosecute those who taught it, were not consistently successful even in the South. Instead, 'Social Darwinism', the application of evolutionary science to sociology and politics most associated with Herbert Spencer and his phrase 'survival of the fittest', resonated loudly during the era of Reconstruction. Spencer's ideas appealed to the intense individualism to which many Americans attributed the country's greatness, and were used to justify the laissez-faire capitalism embraced by industrialists like Andrew Carnegie.[16] Evolutionary cartoons, jokes, songs and advertising appeared in American magazines and newspapers, while the illustrations accompanying a July 1873 article about the original Neanderthal finds in the magazine *Harper's Weekly* are generally considered the first realistic portrayal of an ancient hominid.[17]

Dinosaurs also consolidated their hold on the American idea of prehistory in this era, completing a fascination with the nation's ancient fauna that dated to the republic's earliest days. Plutocrats competed with one another to stock museums with dinosaurs whose overwhelming size, scientific importance and ties to nationalist sentiments matched their own egos. The country's most famous palaeontologists, the Yale professor Othniel Charles Marsh and the self-taught Philadelphian Edward Cope, scoured the west in the so-called 'Bone Wars', unearthing and identifying over 100 species of dinosaurs. Benjamin Waterhouse Hawkins, who had sculpted the Crystal Palace dinosaurs in the 1850s, sensed the new interest and moved to the United States. He was commissioned in 1867 to create similar models for Manhattan's Central Park, only to see the unfinished creatures smashed by Tamamy Hall thugs. Undeterred, Hawkins moved to Philadelphia, where he made a full-size replica of an American dinosaur for the centennial of the Declaration of Independence in 1868. He also designed small dinosaur models and a wall chart of prehistory that were mass-produced in Britain and the United States for decades.[18]

British interest in the missing link shifted into a major key in June 1876 thanks to newspaper reports about a party of German explorers who had docked in Liverpool on their way home from West Africa with a gorilla. No scientifically authenticated live gorilla had yet been seen in Europe. A few weeks later, London newspapers reported that the animal's German owners had named it M'Pongo, after the creature described in the early seventeenth-century by Andrew Battel, an English sailor who had been imprisoned in present-day Angola. Reports in Britain and the United States linked the animal explicitly to the decade-old sensation surrounding Paul Du Chaillu, while newspapers were prone to misnaming Dr Falkenstein, the leader of the party that had captured M'Pongo, as 'Dr Frankenstein', an elision, whether conscious or not, that showed how the gorilla remained a proto-human monster in the popular imagination. By the following spring Pongo – as the animal's name was now commonly written – was being displayed at the Berlin Aquarium alongside his 'cousin', a chimpanzee called Tschego, another name derived from Battel. It would be overreaching to say that the pair performed human roles, but Berliners watched them urbanely drink claret, smoke pipes and play with their pet dog. Superficially, this was simple mimicry, but at a more profound level it reflected the ongoing unease about kinship that had inspired Mr Gorilla and other human apes in the 1860s.[19]

The London Zoo, which already owned several types of great ape, tried unsuccessfully to purchase Pongo.[20] It was more fitting, though, given the gorilla's place in the mass imagination, that the animal arrived in London in July 1877 on a summer-long lease to the Westminster Aquarium, a vast complex of stages, lecture halls and water tanks in the shadow of Big Ben. The 'Aq' had opened the year before as a venue for rational, improving entertainment, but revenues had never met the substantial operating expenses. It had soon been taken over by one of music hall's greatest showmen, the mellifluously named Signor Guillermo Farini – actually William Hunt, a Canadian who had replicated Charles Blondin's feats in walking on a wire across the Niagara Gorge in the 1860s before settling in Britain, where he eventually managed a succession of acrobats and trapeze artists. On retiring from the stage, Farini had traded his figure-hugging costume and Napoleonic whiskers for a frock coat and neatly trimmed beard, achieving a sober appearance that concealed a carnival barker's showmanship and a canny sense of how to exploit the Aquarium's reputation as a serious scientific venue. Farini displayed supposedly edifying exhibits alongside traditional side-show fare, and created the human cannonball by launching a young woman named Zazel from an elastic-loaded tube into a net stretched above the audience.[21]

The onslaught of advertising in which Farini announced Pongo's arrival proclaimed that it was the 'missing link between the human and brute creation', a phrase that Barnum had employed thirty years earlier to describe 'What Is It?'[22] Pongo was accompanied by a man billed as his 'private secretary', a touch that accentuated the animal's supposed proximity to humans and its status as a gentleman.[23] Farini held a private preview for scientists, including Richard Owen, and reporters; the latter laced their descriptions of Pongo with references to earlier gorilla debates like 'ever since the publication of M. Du Chaillu's wonderful stories of the gorilla, he has been regarded as an object of special interest and curiosity; and the London public now have an opportunity of making his acquaintance', underlining the continuing unease about the animal's evolutionary role.[24]

Physical descriptions of Pongo were more stridently racist, thanks to these lingering fears. Racism was expressed in three separate ways. Firstly, it reflected condescending views about Africans. This is evident in a report in the *Leeds Mercury* which claimed that Pongo 'is as like a little Negro boy in the face as a being not absolutely human can be'.[25] The Aquarium's proximity to Parliament and *Punch's* long-standing tradition of depicting Irishmen with simian features also inspired opponents of Charles Stewart Parnell, one of Home Rule's most skilful advocates, to call him 'Pongo', while magazines published myriad jokes and cartoons about Pongo's views on Irish topics.[26] Finally, Pongo had been taken from Africa at a time when the continent was being divided and colonised, inspiring reports that compared gorillas to the Boers, the dour Dutch-speaking farmers who were resisting British expansion in southern Africa, and to the supposedly languid and inscrutable Turks, who had 'retained the features of savage life'.[27] The Ottoman Empire was collapsing, imperilling Britain's access to the Suez Canal and India. Such comparisons trivialised the peoples and cultures that resisted Britain's imperial and foreign policy aims, while helping create vivid public associations that they were less evolved and more 'savage'.

Few people swallowed Farini's claims to have secured the actual missing link. The key difficulty was that Pongo could not interact with audiences as gorilla acrobats or Howard Paul had done. Pongo could not do much more than copy a few human gestures. At some point Farini decided that Pongo might be taught to write. Such an overt sign of intelligence would dramatically prove the animal's consanguinity with humans. But Pongo was an ape. The alphabet was beyond his ken. He simply ate the pencils. The closest he came to fulfilling music hall gorilla fantasies was to swing about his cage on a trapeze. But Pongo mostly sat inertly, oblivious to the people who had paid to gaze on him. He was probably stressed and addled by alcohol. The magazine *Fun* joked that it would be more exciting to see one of his promoters on display – everyone knew about Farini's acrobatic background – while *Punch* published a poetic lament that included the lines 'you may long go / ere you'll meet a sadder creature / duller, drearier / travel-wearier'.[28] Pongo was clearly neither the missing link nor as entertaining as an adept acrobat in a gorilla suit. But a broad swathe of Britain sensed that a remarkable animal was on display, thanks to images, reports and jokes in newspapers and magazines and the photographic postcards that were sold throughout the country.[29] Among the more memorable jokes were those that appeared in *Punch*, in which an articulate Pongo made groan-worthy puns on Latin phrases, Darwin's writings and colloquialisms.[30]

Pongo's visitors included the Prime Minister Benjamin Disraeli, who had an arch, unintellectual approach to the evolutionary debate. He must have known his visit would be spoofed in magazines like *Punch*, which speculated that he had sought Pongo's advice on British expansion into South Africa. The joke was accompanied by a cartoon, in which Pongo appears to have mesmerised Disraeli, suggesting that imperial adventures were illogical and that gorillas remained threatening monsters in the popular imagination. *Fun* took a more militaristic view by jokily ranking the meeting alongside those between Alexander the Great and the philosopher Diogenes and between the Duke of Wellington and General Blücher on the field of Waterloo.[31] Disraeli was followed in mid-August by the Prince and Princess of Wales, who took their twelve-year old son, the future George V, and his eighteen-year-old cousin, the future Kaiser Wilhelm.[32] Reporters in a more iconoclastic, republican age might have pointed out that Pongo had come to London from Germany, much like the royal family. None of Pongo's visitors, notable or not, cared particularly deeply about evolutionary debates; they wanted to be entertained by missing links and human cannonballs without necessarily believing that either was exactly what Farini claimed it to be.

In September, Farini returned Pongo to Berlin, where the animal soon died, causing newspapers and magazines throughout Britain to lament the demise of a creature that had been imprinted in the public imagination.[33] Pongo's departure and death also created opportunities for Farini's rivals to co-opt the interest he had created in this supposed missing link. For instance, in October, the Alexandra Palace in north London introduced a type of baboon with an anaemic claim to being 'rarer than the gorilla'.[34] The tone-deaf phrase expressed a misunderstanding about Pongo's popularity. No other ape, no matter how uncommon, could fill the gorilla's popularly assigned evolutionary role. Moreover, actual primates almost always disappointed because they did not 'perform' as Howard Paul and gorilla acrobats had done.

The most satisfying and successful response to Pongo came from the acrobat Kotaki, who introduced the sketch 'Pongo Redivivus', or 'Pongo reborn', in London in late 1877. Kotaki had come to Britain a decade earlier with the first troupe of Japanese performers to travel abroad as the country opened its borders and Europe became enthralled with its art and culture.[35] It is not clear what Kotaki had done in the years before introducing his new turn, which alluded to Pongo's fate and a very popular, decade-old music hall sketch entitled 'Paganini Redivivus', in which a violinist had enacted scenes from the life of Niccolò Paganini, the short-lived musical superstar.[36] In the new sketch, Kotaki refracted the gorilla acrobats of the 1860s through the lens of an explicitly human missing link. This was clear to the *Era*, which claimed that Darwin had tried 'to show how a monkey can develop into a man, and now, when the controversy is at its height, and the question is not definitely or satisfactorily settled, a gifted and accurate student of the animal world has shown to many an admiring audience how a man can develop into a monkey'.[37] The curtain rose on Kotaki, clad as a gorilla in a cage, out of which he burst in a display of brute strength to clamber around the stage on all fours. He then mimicked an ape's more gentle dexterity by eating an orange that he clasped with his toes, leaping, scrambling and swinging from a trapeze, before assuming an erect and recognisably human posture, removing his mask and bowing to the audience. The mute performance fulfilled popular fantasies about gorillas, while its unmasked conclusion strongly hinted at consanguinity with humans, because, as many people in the audience would have known, by the 1870s a growing number of palaeontologists believed that the missing link would be found in the Far East, as suggested by Kotaki's Japanese features.[38]

Kotaki's salary and the battles over his contract showed that he was far more entertaining than Pongo had been. In January 1878, the owners of Her Majesty's Theatre and the Oxford Music Hall, two of London's biggest venues, sued one another over the exclusive rights to Kotaki's performances. The former testified that he had paid Kotaki £100 for an exclusive agreement, plus £30 per week for three months, with the right to a further three months at £50 per week. There was clearly a lot of money to be earned by booking Kotaki, who must have been one of the best paid artistes of the day. However, he was soon dissatisfied with the arrangement and especially the £25 a week he pocketed after his agent's commission.[39] In April, he unsuccessfully petitioned the High Court to break his contract with his managers, though the judge ruled that its terms did not extend beyond the metropolis. And so over the coming months Kotaki played the biggest London venues, while also extensively touring the provinces.[40]

Kotaki was the most successful of the many artistes who toured Pongo-inspired sketches throughout Britain in the late 1870s and are now only glimpsed in newspapers and magazines. The evidence attests to the immense interest that Farini had generated in missing links. Most of these long-forgotten individuals were probably acrobats who seized on a fleeting recurrence of topical interest, though the craze also swept up the future star Dan Leno, who was already a music hall veteran at the age of seventeen. His family troupe introduced a sketch entitled 'Pongo the monkey' in Rochdale, Lancashire, in February 1878. For at least a year Dan put on a gorilla suit to play Pongo and chase his step-father antically about the stage, striking him repeatedly with a club.[41] The unsophisticated sketch reflected popular ideas about gorillas.

Pantomime, with its emphasis on topical satires, was the natural home for allusions to Pongo. At Christmas 1877, Pongo was conjured by a fairy: a suitable, if macabre way to insert a dead animal into *Saint George and the Dragon* at the Alexandra Palace. Pongo also appeared as an organ grinder's monkey in *Sleeping Beauty* at the Crystal Palace, while Covent Garden's *Puss in Boots* included the veteran minstrel and gymnast Charles Raynor as Pongo, a footman whose 'antics never cease'.[42] Raynor reprised the role the following Easter in the South London Palace's *Jack of Hearts*, in which he and a partner played a game of cribbage with a giant deck of cards bearing images of the British political figures who had just helped to broker an end to the Russo-Turkish War. Popular interest in the event had been captured in G.H. MacDermott's 'War song', which included the most famous line ever uttered on a music hall stage, 'We don't want to fight, but by Jingo if we do!', instantly furnishing the language with a term for strident bellicosity. The card-play never entered the popular imagination, but it alluded equally to the recent diplomatic gambit. The Foreign Secretary Lord Derby, who was at odds with the Prime Minister, was trumped by his eventual successor Lord Salisbury, while the notorious Little Englander, Gladstone, 'falls before the superior merits of Benjamin D[israeli], and the Russian bear has no chance against old John Bull'. Amid this parade of statesmen, 'Pongo is "paired" by Darwin and the player takes sixteen for his "nob".'[43]

Pongo seems out of place to a modern reader, though Victorians would have recognised at least three jokes invoked by his presence. Most directly, a 'nob' is taken in cribbage when a player holds a jack of the same suit as the cut card. Players of the game would have perceived the evolutionary inference that Darwin and Pongo belonged to the same card suit or species. Next, the context of the Russo-Turkish War would have recalled the repeated comparisons that had been made over the previous summer between Pongo and Ottoman officials, while also reminding people that the Prime Minister, Disraeli, the champion of Britain's imperial ambitions, had visited Pongo, and suggested that he, as the overall victor in the pantomime game, had taken the gorilla's advice in diplomatic affairs.

Comparisons between Pongo and African males had never dominated public reactions, though they inspired artistes who played on racial allusions, like Pedro Sterling, a Hispanic-American dancer who had worked for over a decade with British minstrel troupes. In early 1878 he premiered 'Pongo: or the missing link', an updated version of his well-worn sketch about the antics of a pet monkey. At mid-year he renamed it 'Pongo Redivivus', in a direct appropriation of Kotaki's popularity.[44] Sterling knew that Kotaki alone could not satisfy the demand for his sketch, but was confident that audiences would not mind so long as they were entertained. Kotaki, on the other hand, cared very deeply about the appropriation of his artistry and profits. As we have seen, he was not afraid to speak his mind, which he did by placing advertisements in the *Era*, the principal theatrical newspaper, claiming that he was 'the only Pongo'.[45] Sterling replied in the same newspaper with a racist declaration that:

> the individual that calls himself Pongo Redivivus who cautions managers against imitators happens to be a Japanese. Pongo is an animal. I advertise a Pongo Redivivus sketch in which I take the character of Pongo and as far as imitating, I should want far superior than a Japanese to copy from. We

have met once last July in Manchester. He played it one week at the Gaiety; I played it three weeks at the People's. Talk is cheap, but it takes money to buy whiskey.[46]

The fight, which never progressed beyond an exchange of words, demonstrated the profitability of the 'Pongo Redivivus' name, under which both Kotaki and Sterling performed in London, in the provinces and on the Continent until about 1880.[47]

Over the coming years, Pongos and missing links reappeared in pantomimes, and in burlesque versions of popular plays like Edward Bulwer Lytton's melodrama *The Lady of Lyons*, in which a convinced Darwinian declaimed his beliefs in a song that included the verse 'Man's an Anthropoid – he cannot help that, you know / First evoluted from Pongos of old.'[48] A host of Pongo imitators could be seen on provincial stages, including one who appeared under the comprehensive sobriquet 'Pongo, the remarkable man-monkey, Darwin's missing link', while a thoroughbred named Mr Pongo raced with some success in 1882.[49] The ongoing claims by circus owners to have procured gorillas, and advertisements like the one placed in the *Era* in January 1886 selling 'Pongos or Missing Links, or Gorillas or any other name you may call them', show that such animals had caught the popular imagination.[50]

Charles Darwin, the scientist around whom so much of this controversy had swirled, died in April 1882. True to his retiring personality, he had asked to be buried in the local churchyard, but Huxley had other plans. He mobilised influential friends from the overlapping worlds of the Athenaeum Club, the Church, Parliament and the universities to have Darwin interred at Westminster Abbey, an honour reserved for Britain's greatest heroes. Reports published around the world about the funeral provided tangible evidence of a reconciliation between science and religion, while Huxley's role testified to his social, professional and political status. He had matured into the scientist who 'cast a stabilising anchor from the evolutionary ship', by lecturing, publishing and occupying important official posts.[51]

But responses to Darwin's death were not entirely reverential. Farini once again exploited the public mood by introducing an exhibit in December 1882 that he described as a 'talking monkey', 'missing link', 'What Is It?' and 'positive proof of the Darwinian theory' in a scattershot blast of evolutionary allusions.[52] 'Krao', as the new specimen was named, was said by Farini to fill 'a gap which many philosophers before Darwin had seen and endeavoured to account for'.[53] Like P.T. Barnum and all the best humbugging showmen, Farini had no qualms about invoking leading anatomists and evolutionists to support claims that were republished around the world in major centres like Manhattan and Melbourne, and in far more peripheral, though wonderfully named, places like Cheboygan, Michigan and Wagga Wagga in New South Wales, signalling a very widespread fascination with missing links and an understanding of events in London.[54]

The story spread so far and fast in part because Farini wove a tale about Krao's discovery that read like a boys' adventure, with the showman despatching a Norwegian adventurer to search the Burmese jungles for a legendary race of hairy people. He captured a couple of them, but only received King Thebaw's grudging permission to bring Krao to London on condition that Farini adopt her: a subtle accentuation of her humanity.[55] The story interwove evolutionary knowledge with a topical fascination for Burma, a little-known corner of the empire ruled by Thebaw, who had come to power in 1878 after slaughtering eighty-six relatives. The British press had subsequently depicted him as an evil, untrustworthy drunkard, an ideal villain in the narrative about Krao's discovery.[56] At the same time, Farini accentuated Krao's supposedly animal nature by referring to her with an impersonal pronoun, as in his declaration that 'the human intelligence is shown not only by clear speech … but by the readiness with which it learns the meaning and application of words'.[57] This was a far more radical animalisation than showmen had practised on earlier evolutionary freaks, who had been presented as unusual forms of humanity. By contrast, Pongo had been constantly referred to with the humanising personal pronoun 'he', positioning the gorilla as striving to fulfil its evolutionary ascent, whereas Krao, who was self-evidently human, had to be animalised in order to satisfy the conceit that she was a missing link.

As he had done with Pongo, Farini invited reporters and scientists to examine his new 'creature' before she was put on display. They immediately concluded that she was a young girl suffering from an unidentified condition that left her covered in silky black hair, with a disproportionately large head, distended cheeks, extra teeth, vestigial tail and double-jointed hands and feet.[58] Scepticism about Farini's claims drips from reports in newspapers and magazines throughout Britain.[59] And yet Krao was not dismissed outright. Comic magazines were more sympathetic, like *Funny Folks*, which published a spoonerism-filled poem entitled 'The lissing mink'. This included the lines:

With ecstasy my best I cheet,
And fap my sningers too,
For news has come that's swassing peet,
If it be trictly strue.
'Tis said Farini, 'bart' and smold,
Of Barnum's pite the quink,
Has brought within his fowman's shold
That lovely 'Lissing Mink.'[60]

The word-play distorts, without completely obscuring, specific references to Krao. So the poem suggests the ways in which elite knowledge was twisted and refashioned by music hall artistes and showmen for comic effect. Readers understood that the poem played on the concept of the missing link, but the transposed first letters in the term signal that this is all a joke. Similarly light-hearted reactions could be detected in the *Penny Illustrated Paper*, which printed a drawing based on a popular postcard depicting Farini carrying the naked Krao – a discomfiting image for modern viewers. A text underneath asked simply, 'The missing link. Which is it?', exposing the evident deceit.[61] The same furrow was ploughed by the *Illustrated Sporting and Dramatic News*, which published a pencil sketch based on the postcard that depicted Farini winking at the viewer and holding a finger up to the side of his nose, signalling that it was all a big, good-natured joke (see Figure 4).[62]

Most reports simply pointed out that Farini's evolutionary boasts were essential in shaping the way in which people viewed Krao, like the *Illustrated Police News*, which stated that 'if the idea should occur to him [a visitor to Krao] at all of "missing links," he may be excused for feeling that there is still a link to be found – that which will connect Krao with the monkey world'.[63] Crowds clearly took Farini's humbug with the grains of salt it deserved, in part because Krao was very entertaining. She ostensibly played herself, a young girl. She did not perform a pseudo-traditional dance or ritual, as many indigenous peoples had done on British stages, nor did she grunt to demonstrate her kinship with less evolved creatures. It was an easy 'role' for her to play. She was intelligent and personable and greeted visitors with an ever-improving English, an ability to interact that Pongo had signally lacked.[64]

Some scientists, though, interpreted public interest in Krao as lamentable credulity. At the September 1883 meeting of the British Association for the Advancement of Science, the archaeologist J. Park Harrison argued,

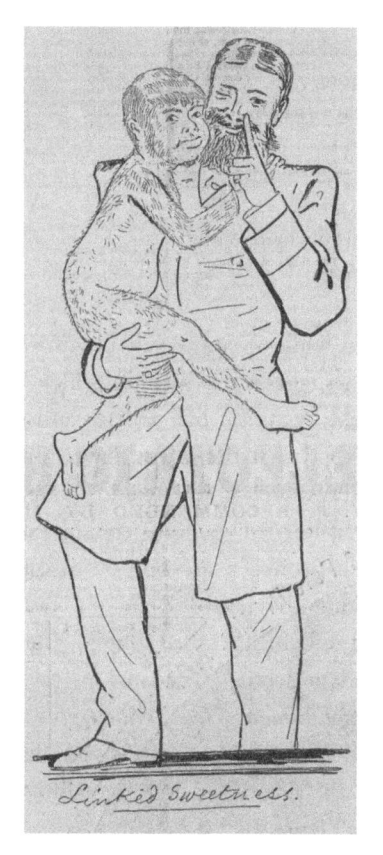

4   An irreverent take on Farini and Krao, *Illustrated Sporting and Dramatic News* (6 January 1883).

without hard evidence, that the number of people who believed Krao was the missing link, an attribution he dismissed as a 'misnomer', made her a legitimate topic for learned debate. The audience laughed audibly when Harrison read excerpts from Farini's evolutionary claims for Krao, and during the ensuing discussions, the chair of the Westminster Aquarium's 'Council of Science', a body whose imprimatur was intended to lend legitimacy to exhibits, stated, unsurprisingly, that it had never been allowed to inspect any of Farini's curiosities.[65] Reports about the debate appeared in newspapers throughout the country, including the *Dundee Courier*, which summarised it bluntly as 'nobody except Mr. Farini and those whom we may call his dupes believe that in Krao we have proof of Darwin's theory of the descent of man'.[66] In other words, eminent minds had proved what had always been obvious to the public.

Once again, scientific truth made no impact on popularity. By the time of the debate, Krao had already been presented to the Prince and Princess of Wales and embarked on a tour of Britain, Europe and North America.[67] Moreover, her departure from London had coincided with the death of Tom Thumb, the dwarf whom Barnum had brought to London in 1844, causing the *Western Daily Press* to state that 'how Barnum missed this catch [Krao] is a puzzle', once again demonstrating a clear understanding of the fake context in which Krao was exhibited.[68] Throughout the tour, Krao was advertised with posters that repeated Farini's evolutionary boasts, while local scientists and physicians were invited to inspect her.[69]

Farini's publicity for Krao was the apotheosis of British evolutionary humbugs and the high-point of his mercurial career. He left the Aquarium in 1885 to travel through Africa, eventually returning to London and increasingly minor managerial roles.[70] In 1894, a greatly reduced Farini admitted that to promote Krao he had imbibed enough Darwinian theory to spin a line of pseudo-scientific nonsense on the pretext that 'who would have gone to see "the hairy girl?" But the "missing link" was quite a different matter.'[71] Farini eventually returned to Canada, where he died in great old age. By contrast, Krao toured extensively in Europe and the United States, where Farini's claims lived on. Audiences may not have believed she was the missing link, but they were charmed and entertained until her death in 1926.[72]

Over the coming years, circuses and travelling waxworks commonly advertised that they had Kraos and missing links. At least some of these were probably created by an east London model maker who specialised in monsters for museums, circuses and theatres, including life-like masks of a 'Burmese hairy family'. He showed an incisive knowledge of Farini's techniques when he opined in 1887 'that a carefully constructed missing link, turned out in the best style, would be a great success, especially if exhibited in the West End, and the public well worked-up previously by a series of scientific paragraphs and quotations from Darwin's works'.[73] Other memorable evocations included a song about Krao entitled 'The missing link, what is it! what is it!! what is it!!!', an increasingly emphatic question that reflected the public debate and drew parallels with Barnum's greatest evolutionary humbug.[74] There were comic lectures on evolution at London's Metropolitan Music Hall, and a comedian toured as 'the missing link', only to be topped by a competitor who billed himself more comprehensively as 'Darwin's missing link'. The latter was also the title of a play performed at Hanley, while a similarly themed production was staged at the Parkhurst Theatre, Holloway, entirely in monkey language.[75] This must have sounded like gibberish, though off stage the term 'missing link' evoked more specific mental images when applied to a boy found living in a wood near Accrington, Lancashire, and when an Irish MP joked on the hustings that his Liberal-Conservative opponent reminded him of Krao, as a missing link between two established political parties.[76]

The story about Krao's discovery in a little-explored part of the world also tapped into the broader idea that missing links might still survive in remote places. These sprang up in the press over the years, such as the report in 1878 that Huxley and Henry Morton Stanley were conferring about whether the latter had finally found evidence of the missing link on one of his African journeys.[77] Equally telling is how the Scottish novelist Robert Louis Stevenson described Jim Hawkins's first glimpses of the hairy and ragged cheese-loving maroon Ben Gunn in the novel *Treasure Island*. In the passage, which was first serialised in *Young Folks* magazine in November 1881, Hawkins sees an indistinct figure moving through the dense tropical forest, wondering 'what it was, whether man, bear or monkey, I could in no wise tell. It seemed dark and shabby', before deciding he has seen a 'lurking non-descript'.[78] The latter term had long been used to claim scientific veracity for evolutionary exhibits.

Lingering interest in missing links had a more direct impact on illustrators, who gradually shifted the context in which ancient hominids were shown from evolution to history. In doing so they began developing a new way of commenting on contemporary society. The roots of this shift can be seen in an 1887 advertisement for Brooke's Monkey Brand soap, which proclaimed that it was 'the missing link in household cleanliness'. The advertisement was surmounted by a drawing of a huge gorilla wielding a frying-pan. A Belfast stationer produced missing-link Christmas cards that year.[79] A brand of bicycles was marketed as the 'missing link', a pun on the links in a chain, while in May 1888 a cartoon in the comic magazine *Moonshine* depicted a man on a tricycle, drawn in such a way as to make him seem simian. It was an apt visual for a cartoon punningly entitled 'The missing link found at last'.[80] But the transition was not abrupt. The Irish continued to be compared to apes, and in 1886 Farini's successors at the Westminster Aquarium promoted a deeply racist exhibition in which a gorilla with pale fur that had been trained to mimic simple human actions was paired with a black one that looked on dumbly.[81] Reports that called the white animal 'half the human race and half the gorilla' played on well-worn fears about consanguinity with apes.[82]

The public largely responded to Farini's evolutionary claims for Pongo and Krao with puns and satires that drew on a very wide knowingness about evolution. Apart from a fairly small group of scientists who cavilled about how the public were being deceived, reactions to these humbugs lacked the vicious tone of the gorilla wars. In those debates, Du Chaillu's account of his African travels had been dissected, but subsequent palaeontological discoveries had created a popular sense that the 'missing link' was an extremely ancient and extinct creature, while the globe was no longer quite so full of unexplored, mysterious and threatening blank spots in which a missing link, be it a gorilla or a girl, might be thought to survive. The popularity of the freak shows that had once been such a staple of British popular culture was also waning. Krao had few successors, as the stage ceased to be a place where evolutionary exhibits derived from the freak show tradition could be displayed. As the remaining chapters show, in the 1890s popular evocations of prehistory ceased being the purview of showmen like Farini who touted missing links or the acrobats who impersonated them on stage. Instead, illustrators, writers and performers explored imaginative spaces within which to satirise contemporary society more directly. In doing so, they created a lasting vision of prehistory as a humorous, archaic mirror of the modern world, peopled by recognisably human characters. The cave man's arrival was imminent.

Notes

1 Thomas Henry Huxley, *Lay Sermons, Addresses and Reviews* (London: Macmillan, 1870), p. viii.
2 Adrian Desmond and James Moore, *Darwin's Sacred Cause: Race, Slavery and the Quest for Human Origins* (London: Allen Lane, 2009), p. 369.
3 Gowan Dawson, *Darwin, Literature and Victorian Respectability* (Cambridge: Cambridge University Press, 2007), p. 5; Alvar Ellegård, *Darwin and the General Reader* (Chicago: University of Chicago Press, 1990), pp. 87–94.
4 The quotation is from Janet Browne, 'Charles Darwin as a celebrity', *Science in Context*, 16:1 (2003), p. 182. See also Janet Browne, 'Darwin in caricature: a study in the popularisation and dissemination of evolution', *Proceedings of the American Philosophical Society*, 145:4 (2001), pp. 496–509; and Constance Areson Clark, '"You are here": missing links, chains of being and the language of cartoons', *Isis*, 100:3 (2009), pp. 573–8.
5 The quotation is from 'Chaldean history of the Deluge', *The Times* (4 December

1872), p. 7. See also M.R.D. Foot (ed.), *The Gladstone Diaries*, vol. VIII (Oxford: Clarendon Press, 1968), p. 252; Irving Finkel, *The Ark before Noah: Decoding the Story of the Flood* (London: Hodder and Stoughton, 2014), pp. 1–7; and Lucy Brown, 'The treatment of the news in mid-Victorian newspapers', *Transactions of the Royal Historical Society*, 27 (1977) p. 34.
6 Glyn Daniel and Colin Renfrew, *The Idea of Prehistory* (Edinburgh: Edinburgh University Press, 1988), p. 41; Ellegård, *Darwin*, pp. 174–215; Louis Figuier, *Primitive Man* (London: Chapman and Hall, 1870); Peter C. Kjærgaard, '"Hurrah for the missing link!": a history of apes, ancestors and a crucial piece of evidence', *Notes and Records: The Royal Society Journal of the History of Science*, 65:1 (2011), pp. 86–9; Ole Klindt-Jensen, 'Dating the earliest Iron Age in Scandinavia', in John D. Evans, Barry Cunliffe and Colin Renfrew (eds.), *Antiquity and Man: Essays in Honour of Glyn Daniel* (London: Thames and Hudson, 1981), pp. 25–7; John Lubbock, *Pre-Historic Times* (London: Williams and Norgate, 1865); Stephanie Moser and Clive Gamble, 'Revolutionary images: the iconic vocabulary for representing human antiquity', in Brian Leigh Molyneux (ed.), *The Cultural Life of Images: Visual Representation in Archaeology* (London: Routledge, 1997), pp. 191–2; Nicholas Ruddick, *The Fire in the Stone: Prehistoric Fiction from Charles Darwin to Jean M. Auel* (Middletown: Wesleyan University Press, 2009), pp. 29–31; Erik Trinkhaus and Pat Shipman, *The Neanderthals: Changing the Image of Mankind* (New York: Alfred A. Knopf, 1993), p. 99–100.
7 'On two recent "scenes" at the Sheffield town council', *Sheffield Daily Telegraph* (18 November 1872), p. 4.
8 See for instance 'The man and the brute', *Glasgow Herald* (11 October 1862), p. 3; and 'A missing link in natural history', *Birmingham Daily Post* (6 February 1865), p. 5.
9 'Judy's half-yearly speeches', *Judy* (26 October 1870), p. 263.
10 'Discoveries for a discoverer', *Punch* (12 October 1872), p. 155.
11 Howard Paul, 'The parents of Adam and Eve: according to Darwin' (Philadelphia: Lee and Walker, 1875), Library of Congress, Microfilm M 3500 M2.3.U6A44.
12 'Mr Howard Paul and the gorilla', *Western Times* (23 April 1873), p. 2.
13 See for instance 'Assembly Rooms', *Cheltenham Looker-On* (21 December 1872), p. 1; 'Music Hall, Worcester', *Worcester Journal* (28 December 1872), p. 5; 'Mr Howard Paul's concert party', *Cheltenham Chronicle* (31 December 1872), p. 5; 'Assembly Room, London Hotel', *Taunton Courier and Western Advertiser* (16 April 1873), p. 4; 'Mr and Mrs Howard Paul at the Mechanics' Institute', *Bradford Observer* (12 May 1873), p. 3; and 'City hall Saturday evening concerts', *Glasgow Herald* (31 January 1874), p. 2.
14 'Covent Garden Theatre', *Morning Post* (30 August 1872), p. 6; and '*Babil and Bijou* at Covent Garden Theatre', *Examiner* (7 September 1872), p. 887.
15 See for instance 'Mr W.B. Fair', *Era* (27 September 1874), p. 16; 'A menagerie under the hammer', *Era* (8 August 1875), p. 3; 'Evening reading', *Isle of Man Times* (12 February 1876), p. 5; 'Les frères Levanti', *Bury and Norwich Post*

(30 May 1876), p. 4; 'Britannia Theatre, Hoxton', *Graphic* (3 June 1876), p. 539; 'Hitcham', *Bury and Norwich Post* (6 June 1876), p. 7; and 'The Middlesex', *Era* (6 August 1876), p. 4.
16 Robert C. Bannister, *Social Darwinism: Science and Myth in Anglo-American Social Thought* (Philadelphia: Temple University Press, 1979), pp. 14–136; Carl N. Degler, *In Search of Human Nature: The Decline and Revival of Darwinism in American Social Thought* (New York: Oxford University Press, 1991), pp. 3–31; James R. Moore, *The Post-Darwinian Controversies: A Study of the Protestant Struggle to Come to Terms with Darwin and Great Britain and America, 1870–1900* (Cambridge: Cambridge University Press, 1979), passim; Ronald L. Numbers, *Darwinism Comes to America* (Cambridge, Massachusetts: Harvard University Press, 1998), pp. 58–75.
17 See for instance 'The Neanderthal man', *Harper's Weekly* (19 July 1873), p. 617; and Stephanie Moser, *Ancestral Images: The Iconography of Human Origins* (Ithaca: Cornell University Press, 1998), p. 136. See also 'Comic zoology: the monkey tribe', *Punchinello* (4 June 1870), p. 150; 'Virginia – a scene in the streets of Richmond', *Frank Leslie's Illustrated Newspaper* (27 April 1872), p. 1; G.B. Bartlett, *Mrs Jarley's Far-Famed Collection of Waxworks* (London: Samuel French, 1873); 'Merchant's gargling oil', Library of Congress, LC-USZ62-75896; and Fred Lyster, 'Evolution, or the Darwinian theory, an anthropological rhyme' (1885), Library of Congress, M 3500 M2.3.U6A44.
18 W.J.T. Mitchell, *The Last Dinosaur Book: The Life and Times of a Cultural Icon* (Chicago: University of Chicago Press, 1998), pp. 131–5 and 156–60; Jane P. Davidson, 'Henry A. Ward, "catalogue of casts of fossils" (1866) and the artistic influence of Benjamin Waterhouse Hawkins on Ward', *Transactions of the Kansas Academy of Science*, 108:3–4 (2005), pp. 144–6; James A. Secord, 'Monsters at the Crystal Palace', in Soraya de Chadarevian and Nick Hopwood (eds.), *Models: The Third Dimension in Science* (Stanford: Stanford University Press, 2004), pp. 138–69.
19 See for instance 'Mail news', *Liverpool Mercury* (22 June 1876), p. 7; 'A strange arrival', *Liverpool Daily Post* (22 June 1876), p. 7; and 'An illustrious visitor', *Morning Post* (16 April 1877), p. 3. Frankenstein appears in 'The gorilla at the Aquarium', *Morning Post* (23 July 1877), p. 6; and 'London amusements', *Chelmsford Chronicle* (27 July 1877), unpaginated. See also 'The educated Berlin gorilla', *New Orleans Daily Democrat* (28 May 1877), p. 2; 'Odds and ends from the old country', *Sydney Morning Herald* (7 July 1877), p. 7; and Colin Groves, 'A history of gorilla taxonomy', in Andrea B. Taylor and Michele L. Goldsmith (eds.), *Gorilla Biology: A Multidisciplinary Perspective* (Cambridge: Cambridge University Press, 2002), p. 15.
20 Untitled, *London Daily News* (25 July 1877), pp. 4–5.
21 Shane Peacock, *The Great Farini: The High-Wire Life of William Hunt* (Toronto: Viking, 1995), pp. 224–86; 'Tight-rope walking over Niagara Falls', *Era* (23 September 1860), p. 13.
22 '"Krao": the missing link', undated programme, British Library, Evanion

76 Inventing the cave man

Collection, 2474; 'Laplanders at the Aquarium', *Era* (18 November 1877), p. 3.
23 'Mr Pongo', *London Daily News* (23 July 1877), p. 1.
24 'A live gorilla in London', *Lloyd's Weekly Newspaper* (22 July 1877), p. 7.
25 'An illustrious visitor', *Leeds Mercury* (14 April 1877), p. 11.
26 'A mischievous crew', *Judy* (8 August 1877), pp. 171–2; untitled, *Leeds Mercury* (10 August 1877), p. 3; 'Mr Pongo on the situation', *Fun* (12 September 1877), p. 111. See also L.P. Curtis, *Apes and Angels: The Irishman in Victorian Caricature* (Washington, DC: Smithsonian Press, 1997), passim.
27 'Pongo', *Examiner* (28 July 1877), p. 11.
28 The quotation is from 'Reflections on the gorilla', *Punch* (4 August 1877), p. 41. See also 'Dangerous rivalry', *Fun* (1 August 1877), p. 51; 'Pongo's memoirs', *Funny Folks* (4 August 1877), p. 34; and 'Pongo the gorilla', *Leisure Hour* (27 October 1877), pp. 686–7.
29 A photograph of Pongo is in the Victoria and Albert Museum collection, E.494-1995; see also 'By private wire', *Fun* (5 September 1877), p. 106.
30 See for instance 'Pongo-isms', *Punch* (18 August 1877), p. 61; 'Pongo-isms', *Punch* (25 August 1877), p. 84; and 'Pongo-isms', *Worcestershire Chronicle* (25 August 1877), p. 3.
31 'Pongo', *Punch* (11 August 1877), p. 58; and 'Does not a meeting like this make amends?', *Fun* (15 August 1877), p. 66.
32 'The gorilla at the Aquarium', *Illustrated London News* (18 August 1877), p. 157.
33 See for instance 'Royal Aquarium', *Era* (2 September 1877), p. 11; 'The death of Pongo', *Edinburgh Evening News* (27 November 1877), p. 4; 'The death of Pongo', *Staffordshire Sentinel & Commercial and General Advertiser* (1 December 1877), p. 2; and 'Pongo's post-mortem', *Punch* (8 December 1877), p. 262.
34 'Alexandra Palace', *Era* (14 October 1877), p. 11.
35 'The Japanese troupe', *Birmingham Journal* (18 May 1867), p. 8; and 'The Japanese troupe', *Liverpool Daily Post* (22 May 1867), p. 4.
36 'Paganini Redivivus', *Era* (7 May 1865), p. 1.
37 '"Pongo" at the Oxford', *Era* (3 February 1878), p. 7.
38 '"Pongo" at the Oxford', *Era* (3 February 1878), p. 7; Kjærgaard, '"Hurrah for the missing link!"', pp. 91–2.
39 'Pongo in Chancery', *Era* (13 January 1878), p.7; untitled, *Theatre* (16 January 1878), p. 392; 'Pongo in Chancery', *Era* (20 January 1878), p. 7; 'Pongo in Chancery', *Era* (14 April 1878), p. 6.
40 'London Pavilion', *Era* (7 April 1878), p. 4; untitled, *Examiner* (13 April 1878), p. 451; 'Alexandra Palace', *Era* (28 April 1878), p. 11; 'The Cambridge', *Era* (19 May 1878), p. 4; 'Manchester – Gaiety', *Era* (14 July 1878), p. 9.
41 J. Hickory Wood, *Dan Leno* (London: Methuen, 1905), p. 18. See also 'Rochdale', *Era* (10 February 1878), p. 9; 'Varieties (Stansfield's) Briggate', *Leeds Times* (27 April 1878), p. 4; and 'Liverpool – Adelphi Theatre', *Era* (4 May 1879), p. 8.

42 The quotation is from 'Puss in Boots at Covent Garden', *Era* (27 January 1878), p. 7. See also 'Alexandra Palace', *Morning Post* (24 December 1877), p. 2; 'Crystal Palace pantomime', *Era* (30 December 1877), p. 13; and 'Mr Fred Johnston', *Era* (6 January 1878), p. 20.

43 'The South London Palace', *Era* (28 April 1878), p. 4.

44 'Christy Minstrels', *Era* (27 December 1868), p. 14; 'Pedro Sterling', *Era* (5 May 1872), p. 15; 'Alexandra Palace', *Era* (28 April 1878), p. 4; 'Sterling, Davis and Sterling', *Era* (10 November 1878), p. 20; Thomas 'Whimsical' Walker, *From Sawdust to Windsor Castle* (London: Stanley Paul, 1922), pp. 110–11.

45 'Pongo Redivivus', *Era* (4 August 1878), p. 13.

46 'To whom the shoe fits', *Era* (18 August 1878), p. 15.

47 See for instance 'Wigan – Alexandra Music Hall', *Era* (15 September 1878), p. 9; 'South London Palace', *Era* (18 May 1879), p. 10; and 'Pongo, Pongo, Pongo', *Era* (1 February 1880), p. 17.

48 The quotation is from William Davenport Adams, *A Book of Burlesque* (London: Henry, 1891), p. 158. See also 'Gaiety Theatre', *Morning Post* (7 October 1878), p. 3; 'The Grecian', *Reynolds's Newspaper* (22 December 1878), p. 5; and 'Alexandra Palace', *Standard* (26 December 1878), p. 6.

49 The quotation is from 'The original Colleen', *Era* (19 December 1880), p. 21. See also 'The Surrey', *London Standard* (27 December 1878), p. 3; 'Poluski, the Little Poluski', *Era* (5 January 1879), p. 15; 'Odd Fellows' Hall', *Era* (1 February 1880), p. 24; 'Manchester Gaiety', *Era* (22 January 1881), p. 10; 'Greenock – Theatre Royal', *Era* (9 April 1881), p. 10; 'Birkenhead – Theatre Royal', *Era* (7 January 1882), p. 6; 'Crystal Palace', *Morning Post* (2 July 1885), p. 1; and 'Birmingham – Concert Hall', *Era* (9 January 1886), p. 14.

50 The quotation is from 'Pongos', *Era* (16 January 1886), p. 20. See also untitled, *Bell's Life in London* (11 March 1882), p. 6; 'Bristol – Sanger's Circus', *Era* (11 December 1886), p. 17; and 'Sport', *Illustrated Weekly News* (21 September 1889), p. 5.

51 Adrian Desmond, *Huxley: From Devil's Disciple to Evolution's High Priest* (Reading: Addison Wesley, 1997), p. xi.

52 The quotations are from 'Krao', *London Daily News* (30 December 1882), p. 4; see also Peacock, *The Great Farini*, p. 288.

53 '"Krao" the missing link', undated programme, British Library, Evanion Collection, Evan.2474, pp. 13–14.

54 See for instance 'Krao the human monkey', *New York Times* (8 February 1883), p. 2; 'Missing link', *Northern Tribune* (Cheboygan, Michigan) (17 February 1883), p. 7; 'A so-called "link" between man and the ape', *Age* (Melbourne) (23 February 1883), p. 3; and 'European intelligence', *Wagga Wagga Advertiser* (24 February 1883), p. 1.

55 'Mr Farini's "missing link"', *Morning Post* (1 January 1883), p. 2; 'Krao, "the missing link" at the Aquarium', *Illustrated Police News* (14 January 1883), p. 2; Nigel Rothfels, 'Aztecs, aborigines, and ape-people: science and freaks in Germany, 1850–1900', in Rosemarie Garland Thomson (ed.), *Freakery:*

Cultural Spectacles of the Extraordinary Body (New York: New York University Press, 1996), pp. 162–5.

56 Deborah Deacon Boyer, 'Picturing the other: images of Burmans in imperial Britain', *Victorian Periodicals Review*, 35:3 (2002), pp. 214–26.

57 'The missing link', *Reynolds Newspaper* (31 December 1882), p. 8.

58 'Krao, the human monkey', *Dundee Courier* (24 January 1883), p. 3; 'The British Association', *Pall Mall Gazette* (26 September 1883), p. 8.

59 See for instance 'Music and the drama', *Glasgow Herald* (1 January 1883), p. 5; 'A "missing link"', *Edinburgh Evening News* (1 January 1883), p. 4; 'The missing link', *Derby Daily Telegraph* (2 January 1883), p. 4; and 'Krao: The missing link', *Lincolnshire Chronicle* (30 January 1883), p. 3.

60 'The lissing mink', *Funny Folks* (13 January 1883), p. 11. See also 'Cuttings from the comic papers', *Tamworth Herald* (12 May 1883), p. 3; 'Varieties', *Exeter and Plymouth Gazette* (29 June 1883), p. 3.

61 Untitled, *Penny Illustrated Paper* (6 January 1883), p. 13.

62 'Our captious critic', *Illustrated Sporting and Dramatic News* (6 January 1883), p. 425.

63 The quotation is from 'Krao, "the missing link" at the Aquarium', *Illustrated Police News* (14 January 1883), p. 2. See also 'London letter', *Western Mail* (8 January 1883), p. 3; and 'Krao: the missing link', *Morning Post* (27 January 1883), p. 3.

64 'Notes of latest news', *Lloyd's Weekly London Newspaper* (31 December 1882), p. 7. See also 'Royal Aquarium', *Pall Mall Gazette* (19 January 1883), p. 13.

65 The quotation is from 'The British Association', *Morning Post* (27 September 1883), p. 2. See also 'The British Association', *Portsmouth Evening News* (25 September 1883), p. 3; and 'The British Association', *Manchester Courier and Lancashire General Advertiser* (27 September 1883), p. 3.

66 'The savants and the missing link', *Dundee Courier* (28 September 1883), p. 4.

67 Peacock, *The Great Farini*, pp. 296–300; 'New Brighton Winter Palace', *Liverpool Mercury* (21 August 1883), p. 6; 'Brighton Aquarium', *Era* (17 November 1883), p. 9; and 'Public amusements', *Liverpool Mercury* (26 April 1886), p. 5.

68 'London letter', *Western Daily Press* (18 July 1883), p. 6.

69 'Our ladies' column', *Preston Chronicle* (17 February 1883), p. 3; '"Krao", the missing link', *Freeman's Journal* (17 September 1883), p. 5; 'Folly', *Manchester Courier and Lancashire General Advertiser* (28 September 1883), p. 1.

70 'Music hall notes', *Pall Mall Gazette* (17 August 1893), p. 4.

71 'A chat with G.A. Farini', *Era* (30 June 1894), p. 14.

72 See for instance 'A puppy boy and a monkey girl', *Seattle Post-Intelligencer* (23 November 1884), p. 1; 'Barton Fair', *Gloucester Citizen* (28 September 1891), p. 3; 'Day's menagerie and museum', *Sheffield Evening Telegraph* (29 December 1891), p. 1; and '"Krao Farini" the missing link', *Chelmsford Chronicle* (1 February 1895), p. 7.

73 The quotation is from 'Monsters and their makers', *Pall Mall Gazette* (26 March 1887), p. 2. See also 'Manders', *Aberdeen Evening Express* (12 August 1884), p. 1;

'Brute biped', *Hampshire Telegraph* (28 September 1889), p. 11; 'Amusements in Brighton', *Era* (25 October 1890), p. 19; 'Reynolds' Exhibition', *Liverpool Mercury* (27 December 1890), p. 5; and 'Amusements in Liverpool', *Era* (28 February 1891), p. 18. A wax effigy of Krao is in the Victoria and Albert Museum, S.484:1 to 3-1979.
74 'Will Lawton', *Era* (3 February 1883), p. 19.
75 'Avenue Theatre', *Sunderland Echo* (3 November 1884), p. 1; 'Cosmo Clarq', *Era* (13 February 1886), p. 23; untitled, *Era* (30 January 1886), p. 10; 'Burton-on-Trent', *Era* (3 July 1886), p. 18; 'Glasgow', *Era* (20 March 1886), p. 17; 'The Metropolitan', *Era* (4 June 1887), p. 16; 'Literary and Philosophical Society', *Leicester Chronicle* (24 January 1885), p. 2; 'Music hall sports', *Era* (9 July 1887), p. 17; 'Hanley – Gaiety Theatre', *Era* (28 April 1888), p. 17; 'Far west', *Era* (26 May 1888), p. 23; 'Felix's', *Era* (23 November 1889), p. 25; and 'Passion's power', *Era* (15 April 1893), p. 4.
76- 'A real "missing link"', *Portsmouth Evening News* (18 January 1883), p. 4; 'Our ladies' column', *Leicester Chronicle* (7 July 1883), p. 3; 'The representation of Limerick city', *Freeman's Journal* (20 October 1883), p. 6.
77 'The missing link', *Edinburgh Evening News* (30 April 1878), p. 4.
78 Robert Louis Stevenson, *Treasure Island* (London: Cassell, 1883), p. 118; 'Treasure island', *Young Folks* (12 November 1881), p. 166.
79 'Brooke's Monkey Brand soap', *Penny Illustrated Paper* (20 June 1887), p. 29; 'Messrs Marcus Ward and Co.'s Christmas cards', *Belfast Newsletter* (28 November 1887), p. 5.
80 'W. Norman', *Taunton Courier and Western Advertiser* (1 October 1884), p. 1; 'The missing link found at last', *Moonshine* (26 May 1888), p. 251.
81 See for instance 'The missing link', *Illustrated Chips* (4 June 1892), p. 8; and untitled, *Funny Folks* (31 December 1892), p. 35; 'The white gorilla at the Westminster Aquarium', *Illustrated Police News* (6 February 1886), p. 1.
82 The quotation is from 'Wanted', *Era* (27 March 1886), p. 19. See also 'A white gorilla', *Standard* (26 January 1886), p. 3; and 'Our extra-special interviews the white gorilla', *Fun* (24 February 1886), p. 79.

4

Antediluvian pictorial fun: E.T. Reed and the prehistoric peeps

By the early 1890s, British newspapers and magazines reported regularly on the discovery of prehistoric hominid fossils throughout the world. The decade's most sensational find was announced in an 1894 monograph by the Dutch palaeontologist Eugène Dubois, in which he described the tooth, femur and skullcap of an extremely ancient bipedal creature that had been unearthed on the Indonesian island of Java. Like his German mentor, Ernst Haeckel, Dubois had become convinced that humans had originated in the Far East, and so he had traded an academic post for that of a doctor in the Dutch East Indian army, believing he would also have time to dig for fossils. Dubois argued that his short, bow-legged, heavy-browed creature was neither entirely ape nor entirely human. It was the missing link. So he named it *Pithecanthropus erectus*, the upright-ape-man, an unsubtle allusion to the pivotal role he believed it had played in human evolution. The creature was more commonly known as 'Java man'.[1] Dubois brought his fossils to Europe in 1895. Haeckel championed him on the Continent, while the young Scottish palaeontologist Arthur Keith led the charge in Britain by proclaiming in lectures and in print that this was the missing link. Thomas Henry Huxley wrote Keith an encouraging letter in which he joked about his own former role as evolution's public champion, stating that the Java fossils made his feud with Sir Richard Owen seem 'positively prehistoric'.[2]

The comedy entitled *The Missing Link* that was staged at London's Surrey Theatre in May 1894 was a bellwether for popular cultural reactions to Dubois's discovery. It was yet another revision of Henry Addison's 1842 farce *The Blue-Faced Baboon*, which had been updated in 1861 as *Mr Gorilla* soon after the cartoon of that name had been published in *Punch*. The basic plot about an ancient hominid invading a middle-class home had been re-envisioned once again in light of sensational and up-to-date scientific discoveries. Equally telling was the fate of P.T. Barnum's mid-century African evolutionary freak 'What Is It?', which toured Britain at the end of the decade as a circus side-show attraction. The role of 'What Is It?' was still played by a black American in a gorilla costume, but he now competed for public attention with the so-called 'Wild men of Borneo', indigenous peoples who were advertised as prehistoric relics from the Far East in a nod to more current scientific knowledge and popular fascinations.[3] Both exhibits showed that freak performers now rarely commanded centre stage as they had once done.

Scientific proclamations that the missing link had been found – especially since they were contested – once again provided British comic magazines with ideal evolutionary fodder. This can be seen in a full-page cartoon in the magazine *Moonshine* in January 1895 showing the Java fossils being unearthed by an Indonesian labourer. They are sold to a scientist, whose chalk drawing of the creature bears a striking resemblance to Charles Darwin, evoking the well-known British debates about human evolution (see Figure 5).[4] At the same time, the imaginative interpretation of the Java bones as those of a Western European shows how the lens through which the British viewed prehistory was being refocussed. The reconstructed creature is not based on Dubois's book, and nor does it resemble the loincloth-wearing Indonesian labourer. The cartoon spoofed the relationship between Darwinian theory and the missing link, and shunted aside fossil evidence to suggest that the creature had been more British than Far Eastern. The image indicated that British cartoonists were reconceptualising prehistory as an era that had been inhabited by people very much like themselves.

It was a short step from portraying the missing link as Darwin's lookalike to transporting the creature to Britain, much as gorilla cartoons had done in the 1860s. A week after the *Moonshine* cartoon, the comic magazine *Judy* published one in which a dandy meets an ugly and ungainly, but not simian, missing link on a London street. The pair get on well over a drink in the pub and then visit a tailor's, from which they emerge clad identically in the most up-to-the minute fashion. This makes them indistinguishable, even to the trained eye of 'Professor Mouldeigh', who looks like Owen, the erstwhile foe of Darwin and Huxley.[5] His dismissive surname and inability to recognise such an important evolutionary relic signalled the triumph of humour over arcane scientific quarrels. The cartoon, and just about all

Figure 5 The discovery of the missing link, *Moonshine* (19 January 1895).

others that followed, reinforced the imaginative, though scientifically invalid, concept that the missing link had been a modern human with Western European features.

Even the most abrupt shifts in popular culture rarely erase earlier fascinations, no matter how outmoded they become. Audiences continue to look for reliable and comforting entertainment from well-worn characters, jokes and sketches. This is evident in the 1890s, because ample room remained in the sparse evolutionary tree for acrobats in gorilla suits to continue billing themselves as 'missing links', 'man-monkeys' and other names that capitalised on residual public interest in gorillas. Their costumes reflected outdated evolutionary ideas, while their tumbling and wire-walking acts had changed little since the 1860s. They now appeared most frequently in grim industrial centres in Yorkshire, Lancashire, Nottinghamshire and Scotland, geographic manifestations of how this simian, Afro-centric, racially threatening conception of evolution had been pushed to the margins of the music hall industry just as it had fallen from scientific favour. It is easy to imagine ageing, aching and stiff-limbed performers in threadbare, patched and time-worn costumes earning small wages while attempting to please audiences with unsophisticated tastes. Such was the fate of artistes who could not update their turns.[6]

Few such performers appeared in London. Metropolitan music hall owners could select the most sensational and topical acts from among the country's largest pool of talent. Moreover, by the early 1890s, music halls had largely been incorporated into regional and national syndicates that focussed on attracting middle-class ticket buyers. The term 'music hall' was gradually being replaced by the more genteel-sounding 'variety entertainment', even if performances could still be raucous and risqué, and topicality remained as important as ever. The social aspirations of the men who controlled the industry left little room for gorilla acrobats or sketches derived from the freak tradition.

The use of the term 'missing link' in popular culture was also shifting. It was employed in casual speech to describe odd-looking or unattractive people, while participants clad as missing links, often riding machines that had been made to look archaic, regularly participated in the cycling parades that proliferated in Britain in the 1890s.[7] The latter reflected how the globe's fairly comprehensive exploration and colonisation prompted people to abandon simplistic notions that missing links had survived in remote regions. Instead, northern Europe's prehistory was becoming much more commonly understood thanks to excavations of barrows, hill forts, henges and other sites throughout the Continent. In Britain, this emerging knowledge aligned with the Victorians' wider romanticisation of the nation's past, through such things as the Arthurian poems of Alfred, Lord Tennyson, Gothic architecture and Pre-Raphaelite paintings. These provided comforting, if unscientific, visions of an eternal Britain, whose inhabitants had outlived all other hominids and now topped a hierarchy of races.

Over the past century and a half, Impressionism, Fauvism, Cubism and many other 'isms' have radically altered perspective in painting; architects have abandoned stone trefoils and ogees for glass and steel; and poems adhere to new notions of metre and rhyme. And yet the cave man that emerged in British pictorial and literary manifestations of prehistory in the 1890s is largely unaltered today. This is because the malleable imaginative space in which writers and illustrators worked allowed them to conjure up a prehistoric world that mimicked late Victorian Britain with recognisable, if archaic, institutions, technology, family life and middle-class notions of civility. As they did so, gentler comic commentaries on current events supplanted the farces and shattering sexual and racial threats that gorillas had once represented, and were unrelated to the freak show tradition on which Farini had drawn for his missing links. Ever since, cave men, depicted with remarkable consistency, have populated comic spoofs set in barely disguised versions of the contemporary world.

The shift, which slightly predated the discovery of Java man, can be illustrated by juxtaposing two literary evocations of prehistory that appeared in 1889. That year, Samuel Page Widnall, a prosperous but eccentric Cambridgeshire farmer, published *A Mystery of Sixty Centuries, or a Modern St. George and the Dragon*. It is a quotidian boys' adventure about three public-school chums exploring the African interior. They meet indigenous people who tell them about a race of flying dragons, which the protagonists initially believe are the Biblical Nephilim – antediluvian giants – before realising they are pterodactyls. It is a ham-handed attempt to reconcile evolution, Biblical revelation and British history, as the titular reference to England's patron saint suggests. This self-published novel lacked literary merit, and reflected outdated evolutionary debates and a long-surpassed sense that prehistoric creatures had survived in the unexplored parts of the world. So it is unsurprising that *A Mystery of Sixty Centuries* was not widely circulated, reviewed or collected by depository libraries.[8]

Henry Duff Traill's 'A day with primeval man', which appeared in the March 1889 issue of the *Universal Review*, was a much more important, more topical and better-written evocation of prehistory. It is the story of a man who naps on Wimbledon Common and dreams of the area in prehistoric times. He watches hunters stalk elk and bear across a glacial landscape, while their women and children wait in a hut – familiar nineteenth-century gender roles. Traill, an Oxford-educated journalist, literary critic and biographer, crafted a story with sufficient excitement to divert readers, without worrying too much about scientific accuracy. At the same time, he subverted heretofore central elements in literary manifestations of prehistory by locating the story in Britain rather than a distant, exotic and little-known part of the world. He may well have borrowed the structure from the French scientist Pierre Boitard's 1861 work *Paris avant les hommes*, in which the protagonist takes a fantastic dream voyage to a prehistoric France inhabited by orang-utan-like hominids.[9]

By contrast, Traill described his creatures simply but tellingly as 'humans' and 'sinewy', reflecting more recent scientific discoveries and images of Cro-Magnons in popular works like Louis Figuier's *Primitive Man*.[10] Traill's description freed Charles Shannon, one of the greatest Victorian illustrators, to transpose modern humans into a prehistoric London landscape for the nine drawings that accompanied the story (see Figure 6). Shannon laid bare his artistic training. The animal skins and beards in which he clad the figures reflected canonical motifs for depicting wild men and Classical heroes, while their lissom bodies were indebted to the Pre-Raphaelites, and the mountainous landscape and wooden houses on stilts were inspired by the Japanese artist Hiroshige, who was revered by Shannon's generation. As such, the images fundamentally break from the populist work of Boitard's and Figuier's illustrators, and from French academic traditions of depicting the deep past with archaeological accuracy.[11]

Shannon was a commercial artist hired to create illustrations that reflected a writer's ideas. As it was for many others who worked on publications aimed at broad audiences, it was more important for Shannon to be entertaining than scientifically accurate. And yet images have been endowed, since at least the development of perspective in the Renaissance, with the power to convey important messages.[12] People learned two significant things from prehistoric cartoons in the 1890s that have never since been dislodged from the popular imagination. Firstly, prehistory is an archaic reflection of the contemporary world. This made it primarily a historical rather than an evolutionary era, because audiences were asked to believe that technology and institutions had become more refined and complex over time, but that Western Europeans were unchanged. The first gorilla cartoons in *Punch* had played on the deeply threatening notion that the British were descended from African apes. Cartoonists now suggested that this was not so and that the British were eternal. This gave

A PRIMEVAL BATHING-PLACE. C. H. Shannon.

6   Charles Shannon, prehistoric Wimbledon, *Universal Review* (March 1889).

subtle credence to ideas of national greatness and implied that other races and peoples were destined to remain unevolved and primitive. Secondly, prehistory became inherently comic. It related only tangentially to palaeontological evidence, while the lack of written records made it well-nigh impossible to draw moral, political or imperial lessons from the era as the Victorians did from Classical texts. Prehistory could be used to spoof and shore up national ideals, but it lacked *gravitas*.

Three drawings that appeared in *Punch's Almanack* in December 1893 cemented prehistory as a humorously archaic version of contemporary Britain. One depicted a pair of cave men at a stone billiards table over the words 'Primeval billiards'; a second featured a London cab made of rough timbers with solid wood wheels being drawn by a bear across boulder-strewn ground identified as 'The first hansom'; and the third showed a furious fight over a round of cards, or as its subtitle stated, 'A little quiet whist in prehistoric times – The end of the game.'[13] A fourth one, depicting a group of hunters being pursued along a narrow mountain path by a woolly mammoth appeared in the magazine a few weeks later. The characters in this cartoon leap into the abyss, cower behind boulders or careen headlong towards the viewer.[14] Three of the cartoons bore the title 'Prehistoric peeps', by which the series would be known. The phrase made *Punch*'s satirical intentions clear by simultaneously invoking the verb to 'peep', or spy quickly, voyeuristic peep shows and the seventeenth-century diarist Samuel Pepys, slyly suggesting that the cartoons were accurate glimpses of a far-distant past. Long-standing *Punch* readers would also have recalled an allusion to 'Peeps at Paris', a series of tongue-in-cheek articles describing life in 'Parry' that the magazine had published in 1867.[15]

Three of the images commented on contemporary society: the hansom cab and billiards table are fancifully archaic versions of modern contraptions, while whist was widely played amid the aspidistras and gewgaws of middle-class drawing rooms. The knockabout violence and farce that would fill virtually every one of the forty or so subsequent peeps drawings sent up Victorian codes of decorum, sportsmanship and civility. The inhabitants of this world, known forever after as 'peeps', are slight, awkward, unkempt and recognisably human. Their bearskin costumes and stone axes reflect Western traditions for depicting wild-men, while their hair, gestures and mugging faces are rooted in *Punch*'s comic aesthetic. The humorous conceit underlying the cartoons required *Punch* readers to recognise themselves in these scruffy and socially maladroit characters, which would have been impossible if they had resembled gorillas, Pongo, Krao or any other stand-in for the missing link.

The cartoons were sensationally popular from the start and remained so twenty-five years later, an astonishingly unquenchable public infatuation.

Their immediate reception recalled the one that had greeted 'Mr Gorilla' in 1861. That had been a single image, whereas cartoon after cartoon in which quasi-modern humans inhabited a recognisably archaic version of contemporary Britain reinforced a particular imaginative view of the ancient past. The cartoons were all drawn by Edward Tennyson 'E.T.' Reed, a Harrow-educated artist who had studied with the Pre-Raphaelite painter Edward Burne-Jones before becoming a commercial illustrator. In 1889, a friend suggested that Reed submit comic drawings to *Punch*. One of these, depicting a human skeleton mounted in a museum display case, was published that July.[16] The creature's dramatically hunched back and tiny, almost vestigial arms sit atop a sturdy pelvis, massive, elongated leg bones and claw-like feet. It was a satire on the craze for bicycling, depicting how people might evolve if they continued pursuing the sport. Reed titled it simply 'A warning to enthusiasts'. It was not the first cartoon to make evolutionary jokes about cyclists, but Reed gave his image a deliberately scientific slant, telling an interviewer that he wanted to show how such a skeleton might be unearthed and exhibited 'a thousand years hence', much as palaeontological specimens were then being displayed to the British public.[17]

On the basis of these drawings, Reed was invited to join the staff of *Punch*, after which he continued producing cartoons that satirised evolution and prehistory. Two of these showed his personal familiarity with how scientists and showmen had promoted evolutionary ideas over the previous decades, and strongly suggested that *Punch* readers were expected to be equally conversant. The first was 'Evolutionary assimilation', a July 1890 cartoon featuring five images of the Italian cellist Carlo Piatti, who had recently performed in London. At the far left, Piatti appears as a white-haired man in pince-nez bowing the strings of his instrument. As the images progress to the right, he 'evolves' by increasingly taking the shape of his instrument, until the final one, which shows only two intertwined cellos.[18] Readers would have recognised the obvious allusion to the evolutionary illustration that Benjamin Waterhouse Hawkins had created for Huxley's 1863 book *Evidence as to Man's Place in Nature*. Reed followed this in September with 'Professor Marsh's primeval troupe', which satirised the widely reported presentation that the Yale palaeontologist Othniel Charles Marsh had just made to the British Association for the Advancement of Science. Marsh had unveiled dinosaurs from the American west, causing Reed to depict him as a Yankee showman in a garish tailcoat emblazoned with stars and stripes. He stands on a lecture stage among prehistoric creatures – the first dinosaurs that Reed drew – that are posed like circus animals. The image likened Marsh to Barnum and other humbugging showmen who had toured Britain with supposedly ancient and evolutionary specimens over the previous half century.[19]

In March 1894, Reed was appointed parliamentary cartoonist, *Punch's* most prestigious post. Thereafter, his drawings of political figures, either alone or in groups, as traditional character studies or in elaborate fantasies, appeared in virtually every issue. They constitute the vast majority of his output, though he never stopped publishing cartoons that reflected his interest in the distant past. The two streams merged at times, notably when Reed drew peeps cartoons featuring recognisable caricatures of Victorian and Edwardian personalities. For instance, he celebrated his appointment as parliamentary cartoonist with a drawing depicting budget night in a prehistoric House of Commons. A Speaker cradling a giant stone axe oversees the chamber, his authority represented by the wooden club that lies atop the table where the mace would normally rest. Many of the peeps who crowd the benches are recognisable as sitting parliamentarians, such as the Chancellor of the Exchequer, Sir William Harcourt, who reads the budget speech from a stone despatch box, a marmoreal William Ewart Gladstone and a preening Joseph Chamberlain. Lest anyone forget that this is a Stone Age scene, three long-necked dinosaurs peer over the chamber's high rock walls.[20] The image departed radically from earlier ones in which hominids had been depicted hunting ancient elks, bears and mammoths in fairly realistic settings. But Reed placed modern humans alongside dinosaurs that had died out millions of years ago, signalling that his entirely whimsical world could not be constrained by scientific evidence. The era was a platform from which to satirise contemporary Britain. It was a very comforting conceit because it conveyed the message that British institutions and gentlemanly ideals were immutable and universal, and had always been at the apex of civilisation, thereby subtly validating imperial aspirations.

Reed was cagey about where the idea came from, suggesting in one interview that he had been frustrated in trying to draw a series of cartoons set thousands of years in the future, only to be inspired by the dinosaurs in the Natural History Museum. In another, he claimed that the peeps had been 'created on the spur of the moment. I found the "new line" was favoured by the public, and I like looking into the far distant past. What *we* [sic] might have been.'[21] But the reality was less damascene. Reed had been born in 1860, the year the Darwinian debate had exploded in Britain, and had grown up amid increasing public interest in prehistory, evolution and missing links. His Classical education at Harrow had been designed to inculcate gentlemanly values, while his formal art training introduced him to Western traditions for depicting the ancient past. He also kept prehistoric relics in his studio and read books on the topic, but insisted on the caveat that his creatures were 'real animals – reconstructed from fossilised bones. I get them out of palaeontological books. They are quite genuine – with just a touch of exaggeration.'[22] Reed could just as easily have been referring to his parliamentary caricatures, showing that they all reflected *Punch*'s satirical tone.

Prehistory was an almost ideal vehicle for satire, being sufficiently familiar to *Punch* readers for them to recognise cues and allusions, while also distant enough that they could laugh at themselves and their beliefs unselfconsciously. A series set in the far future such as Reed had first tried to create could not serve this function because of its entirely speculative nature. Science fiction has frequently been used to teach moral or political lessons about the present day in ways that prehistory cannot do convincingly. The humour in Reed's sole futuristic image, his warning of the evolutionary consequences of bicycling, is actually a projection into the future of what readers believe to have happened in the prehistoric past. Reed clearly perceived that prehistory was sufficiently malleable that almost anything could be included – witness the intermingling of humans and dinosaurs – to support his flights of fancy. Reed's prehistory was also fundamentally conservative and reassuring because no matter how chaotically he depicted this earliest British era, the cartoons always fundamentally proved the immutability of national ideals and institutions.

Little is known about Reed's early output, though a curious forerunner of the peeps appeared in 1888 in the *Boy's Comic Journal*, an illustrated magazine aimed at working-class youths. The unsigned full-page satire on prehistoric courtship may very well have been drawn by Reed, given that the artist's pencil strokes, the men's and women's gangly bodies, clubs, clothing and the comically pop-eyed dinosaurs strongly resemble the ones he would use extensively in the peeps.[23] Even more intriguing is that a slightly altered version was reprinted in August 1890 in the Anglophile New York comic magazine *Puck*. This cartoon contained two additional scenes, and bore the signature of Frederick Burr Opper, who had already drawn humorous gorillas in the style of *Punch* and was becoming one of the most important comic illustrators in the United States.[24] The figures in this cartoon differ very significantly from Opper's other cave men, which are discussed in the next chapter, suggesting that he may have appropriated Reed's work as his own.

Radical differences between Reed's prehistoric Britain and the one that Shannon had drawn belied the similarity of their artistic training. But unlike Shannon, Reed had complete control over his prehistoric world, which was first and foremost a visual one. Moreover, *Punch* was Britain's most widely circulated magazine, while its well-established and easily recognisable aesthetic and satirical spirit heavily influenced cartoon art around the world. Reed differentiated the peeps from other works he produced for the magazine, while also eschewing contemporary artistic trends or scientific rigour. He built up his prehistoric figures with strong, dark pen strokes, left large parts of many images untouched and made minimal use of shading or cross hatching. In spirit, Reed was a classic *Punch* artist who used cartoons to satirise the current day. In order to do so, he took Britain's prehistoric past for granted, transposing contemporary institutions and events into a long-ago setting. The result was a male-dominated, conservative world in which late Victorian assumptions about gender, race and class were replicated unquestioningly. At the same time, though individual peeps sometimes bear the faces of well-known political figures or sportsmen – demonstrating how fluidly Reed moved between parliamentary and prehistoric work – there are few signs like fashion or furnishings that tie the images concretely to a particular period. Someone stumbling on one of the peeps cartoons today would find it hard to date accurately, in part because of the way in which Reed's vision of prehistory has been replicated time and again for more than a century.

Within a year of the first peeps cartoon, Reed had relegated those featuring simian hominids to museums and serious scientific texts. Reactions to the first peeps cartoons demonstrated Reed's innovative take on prehistory. The *Morning Post* newspaper crowed, 'there is perhaps nothing happier, albeit its humour is of the simplest, and its art of the least elaborate, than Mr Reed's "Prehistoric peeps"', while over the coming months newspapers and magazines throughout Britain analysed and applauded each new image and published imitative cartoons.[25]

Reed also incorporated into his images broad music hall devices that would have been obvious to *Punch* readers. As Peter Bailey has shown, artistes alluded to common knowledge and used simple costumes, broad gestures and unsubtle expressions to forge an attentive 'audience' out of a 'crowd' that was constantly distracted.[26] Reed satirised many of the same events, institutions and pastimes as the artistes, arranging his images in a quasi-theatrical way on a flat plane against a simple backdrop, with entrances and exits on either side. The poses of the rangy characters suggest fast movement and farce, in sharp contrast to Reed's parliamentary caricatures, which are generally more static. These elements are evident in many peeps drawings, though the most direct allusions to the halls are a pair of images set in theatres that show dinosaurs attacking the actors on stage.[27]

Reed's desire to depict Britain's ancient past is evident in the fact that only one of his cartoons directly evoked Darwinian theories. This was a June 1894 cartoon entitled 'A night lecture on evolution', showing a group of adults sitting on stone benches as they listen to an impassioned lecturer (see Figure 7). It is a complex image with a unique dark and brooding style that conveys its singular social, political and religious commentary. The furry creatures' long tails and pointed ears allude to arguments Darwin made in the 1870s about vestigial simian features. These imply that the working classes, at whom such night lectures were directed, were less evolved than the middle-class characters Reed normally drew.[28] The disparity is reinforced by the way in which Reed made the impassioned speaker resemble Thomas Henry Huxley, who had lectured regularly on evolution to working men since the 1860s, making himself the target of earlier satires.[29] He leans over a pulpit rail in a sermonising posture, while the tapers and pew-like benches show that the scene takes place in a church. However, the lectern on which a presumably scriptural tome rests is formed by the outstretched wings of a pterodactyl, rather than the eagle that is commonly found in Anglican churches. Portraying Huxley as a blood-and-thunder preacher deflated his agnosticism and attempts to make science the new orthodoxy. At the same time, the bored, yawning and dozing audience guyed the effectiveness of his do-gooding.

Several of Reed's early *Punch* drawings had depicted black people as thick-lipped semi-simians, betraying prevalent racist assumptions, but he only once made these the focus of the peeps. This was because the series was intended to satirise British institutions and events that were largely restricted to the ethnically homogenous upper middle classes.[30] Moreover, the Afro-centric concept of evolution that had predominated in the 1860s had been displaced by an imagined prehistoric Europe. As a result, the only peep cartoon that depicted Africans, and the only one set outside Britain, was 'Prehistoric Fashoda', which appeared in October 1898 during the heated Anglo-French contest for control of a small fort at the headwaters of

![E.T. Reed, a night lecture on evolution, Punch (23 June 1894).]

E.T. Reed, a night lecture on evolution, *Punch* (23 June 1894).   7

the White Nile (see Figure 8). Reed drew the six-member garrison as indolent, preening and overweight, reflecting long-standing British prejudices about the French. They are aided by eight African soldiers depicted with the wide-mouthed, thick-lipped grimaces of music hall minstrels, and armed with clubs and spears that signal a lower place on the evolutionary scale. Without denying or diminishing the crude racism, it is important to point out that Reed was also immensely condescending to the French soldiers. Drawings such as this reinforced dismissive attitudes towards indigenous peoples and Britain's European rivals. As such, the cartoon reflects the unquestioned sense of superiority of a British man whose father had been knighted after a varied career as a naval architect, railway owner and Lord of the Treasury. Reed also embodied this gentlemanly ideal. His neat beard and moustache, frock coat, topper, kid gloves and walking stick made him look remarkably like the future King George V.[31]

Reed's wealth, education and position at *Punch* placed him firmly in the imperial elite, and gave him the confidence to consider every British institution, craze and creed as a legitimate target. So it is impossible to identify a consistent political or moral stance in his peeps drawings apart from a self-satisfied, whimsical view of the middle classes, from the dully aspirational inhabitants of Pooterdom to the secure sense of place felt by those, like Reed, who had attended major public schools and Oxbridge, or occupied important state and institutional offices. Reed cast a broad net

8   E.T. Reed, prehistoric Fashoda, *Punch* (15 October 1898).

over the coming years, drawing cartoons that satirised the cult of games – rugby, rowing, golf and cricket; aesthetic and artistic pretentions – the Royal Academy and the art market; leisure – seaside holidays, charabanc excursions and tea dances; state institutions – Parliament, the Royal Navy and the Lord Mayor's show; and more mundane crazes like ping-pong. When the subjects he drew had obvious aristocratic associations, as in the case of thoroughbred racing and yachting, he portrayed them through a quasi-democratic lens like the Epsom Derby or the America's Cup race.[32]

Political figures were an especial target throughout, most notably the thrusting Joseph Chamberlain, who was often depicted by Reed in the peeps and elsewhere as a long-necked dinosaur. At times, Reed's subjects handed him cloth wholesale. The future Prime Minister Henry Campbell-Bannerman set himself up for satire by declaring at a London banquet in April 1901 that when it came to financing Britain's military, he and his close ally Lord Edmond Fitzmaurice were throwbacks to the Liberal Party's sceptical, anti-imperial 'prehistoric' age. Within the week, Reed drew the pair as cave men, giving the lean and angular Fitzmaurice the pointed simian ears of an early hominid, while the corpulent Campbell-Bannerman rests his arms on his impressive belly in a self-satisfied pose. The fence on which the pair are seated and the handkerchief suspended from the latter's stone axe, which suggests it is an unused adornment to his outfit, show their propensity for vacillation over military action. The title describes them punningly as 'Survivors of the (glad-)stone age'.[33]

Since the 1840s' *Punch*'s editors had regularly published collections of cartoons in order to expand the magazine's audience and generate profits.[34] Their approach to Reed's work was no different. Twenty-six peeps cartoons, some of which had never appeared in the magazine, were issued in a handsome volume in November 1896.[35] A deluxe edition of the book appeared at Christmas two years later, along with *Mr Punch's Animal Land*, in which Reed reimagined prominent people as prehistoric animals. Reed drew himself in the last sketch as a creature which he described in prehistorically ungrammatical language as 'The Wunnudiddit', a beast with the body of a scaly kangaroo, described as 'the anommylous auther of this Ceres. He got all in among the Stone Age once and kept on doing the most ebsurd picktures. He is a kind of Preestorick Pepys. They were a ruff lot according to him they occupied all there spare time chopping oneynother up and dodging the most loathsome lumpy Animals.'[36] It was a close to a credo as he ever came. Three more prehistoric drawings appeared in *Punch's Holiday Book* in June 1901, which also included Reed's tongue-in-cheek confession that his 'feeling for physical beauty of an unheard-of kind' combined a close study of Classical works with 'the influence of Sir Daniel Leno, [and] Sir Gustavus Elen', two of the greatest music hall comedians, showing how casually he melded elite and mass cultures.[37]

Equally telling is how Reed collaborated with the Fine Art Society, one of London's most important commercial galleries, in 1899, and with the Swiss-born engraver François Lutiger to create *bas reliefs* of 'Primeval billiards' and 'The first hansom'.[38] This recreation of five-year-old images demonstrated the unrelenting hold that Reed's vision of prehistory had taken on the public's imagination. Lutiger produced twenty-five silvered-copper, oak-framed versions of each image. They were sold at the society's Bond Street gallery in conjunction with an exhibition of Reed's drawings that opened at the end of June with a private viewing attended by many of the political and literary figures whom he had guyed in prehistoric and parliamentary cartoons. The show received generally positive reviews in newspapers throughout Britain.[39]

Unlike many cartoonists who published simultaneously in several magazines, Reed was bound to *Punch* alone. In mid-1894, when the *Idler* asked him to submit prehistoric drawings, *Punch*'s proprietor insisted that the peeps had 'added so much to the attractiveness of *Punch* for some months past, and are now in the eyes of the public associated with *Punch* and *Punch* alone, we cannot help feeling very much adverse to your drawing anything of a similar nature for another paper'.[40] It would be several years before Reed published prehistoric cartoons in other magazines.[41] Despite shackling Reed, *Punch*'s owners felt no qualms about publishing cartoons that were heavily derivative of the peeps. For instance, in 1897, Charles Harrison began drawing a series set in ancient Egypt, Greece and Rome. Reed also tried his hand at humorous depictions of more recent events from Britain's past under the title 'Unrecorded history' as well as Babylonian-themed cartoons called 'The tablets of azit-tigleth-miphansi, the scribe'. The latter were full-page satires that featured friezes based on the monumental Assyrian sculptures in the British Museum, coupled with elaborate jokes written by Reed and presented in faux-archaic language and script.[42] No other era matched prehistory's popularity; either they were associated with foreign lands and so could not be as easily exploited to comment on modern Britain, or public knowledge about them was more rigidly delimited by the effects of school texts and sober museum displays.

Reed's income from *Punch* reflected the popularity of his drawings. In 1893, at the end of which the first peeps appeared, he earned roughly £300, equivalent to the pay of a senior clerk, but by 1897 this had risen to more than £1,000 per year, or about the income of a barrister or top civil servant. His material circumstances changed accordingly. Reed first drew for *Punch* while living in West Kensington, an upmarket area inhabited by many artists, but by the turn of the century he lived in a detached house overlooking Wimbledon Common, a much grander and more socially exclusive address.[43] He also became an increasingly public figure, a status he evidently enjoyed and used to promote his work, beginning with the telegraphic address he chose for his home, which was simply 'Prehistoric'.[44] He was a genial man, known to his friends as Ted, with an unthreatening sense of humour.[45] By the turn of the century he was giving illustrated lectures about the peeps and parliamentary caricatures to middle-class literary, historical and scientific societies from the Home Counties to Aberdeen and Bristol, events that were reported on as far away as Australia. Invitations to speak to such groups reflected how the general social status of cartoonists had risen over the course of the nineteenth century, while Reed was unquestionably a gentleman, a status that was confirmed when the Viceroy of India requested that he work as an official artist for the 1903 Coronation Durbar.[46]

As with earlier evolutionary sensations, casual and easily understood allusions to the peeps could soon be heard in idiomatic speech throughout Britain, a rich and comprehensive incorporation into daily life that reflected widespread knowledge of Reed's work. Scientists illustrated lectures with Reed's cartoons, while newspapers and magazines invoked the peeps to make reports about zoology and archaeology more digestible.[47] Many of these references compared modern Britain to its archaic predecessor, demonstrating what people found so attractive in Reed's work. For instance, London cabbies hailed one another as peeps in reference to the hansom depicted in one of the first drawings, Cardiff whippet races seemed so archaic as to have been conjured by Reed, and public speakers used the term 'peeps' as a comic shorthand for self-flattering anecdotes about how far people and institutions had evolved over time.[48] Newspapers reported on references to the peeps in a Glasgow courtroom, while shortly after Reed spoke in Sheffield, a local alderman was called a prehistoric peep for his antiquated views.[49] An 1898 report about an expedition to the Kamchatka Peninsula described native boats as something Reed 'could hardly have failed to recognise in the originals what he had only drawn from imagination', showing how the cartoons had become an easily understood way to simultaneously denigrate foreign cultures and underline British exceptionalism.[50] Reed also provided a cosy way of conceptualising ancient Britain, like the day trips from London to Stonehenge that the travel agency Thomas Cook and the Great Western Railway jointly launched in 1899. These were commonly referred to as 'Lunch and prehistoric peeps'.[51]

Such examples demonstrate that British people identified themselves, their institutions and their society with the world Reed drew.

Punch's reach and influence also made the peeps an international phenomenon. Individual cartoons were discussed and reprinted almost immediately in the United States, where 'Primeval billiards', one of the first ever published, was used in an advertisement.[52] Imitative cartoons, printed under the heading 'Prehistoric peeps', appeared in newspapers as far afield as Salt Lake City, Utah, a little over a year after Reed's first drawing in *Punch*.[53] By then, Reed's cartoons were so well known that an obituary of the popular prehistoric writer Louis Figuier, which compared his works to the 'Prehistoric Peeps of *Punch*', was reprinted in newspapers from Ohio to Hawaii.[54] In New Zealand, schoolboys cheering raucously at an 1895 swimming championship were described as peeps, and lantern slides of Reed's cartoons were shown at the meeting of a camera club.[55] Such references attest to the many ways in which Reed's images broadcast the eternal nature of British society, and by extension Anglo-Saxon society, that appealed throughout the English-speaking world. However, the United States and New Zealand had sizeable and ancient indigenous populations that were never depicted. In such countries and colonies, Reed's cartoons propped up the settler people's self-identification with Western Europe, rather than the lands in which they lived.

One of the most startling reactions to the peeps, and one that shows how deeply British people identified with them, was seen in the way they were recreated in settings that evoked Reed's archaic ones. This began in August 1894 when the members of the Molesey Boat Club on the Thames leavened their annual regatta by enacting a cartoon that had appeared in *Punch* the previous month. This had depicted a prehistoric Henley Regatta, another of the most important rowing competitions on the river. Reed had drawn a chaotic scene of peeps paddling coracles, punts, canoes, skiffs and animal-shaped boats on a river teeming with predatory monsters. One bank is lined with barges and houseboats like the ones used by regatta spectators, while the other is filled with tepees from the North American plains (see Figure 9).[56] The cartoon inspired the Molesey club's members to perform a sketch that the programme described in faux-archaic phrases as 'a contemplation of ye ancient Britons, at home on wooden rafts, besmeared with woad in lieu of garments, their ways strange, their customs curious – a sight for ye Gallery Gods'.[57] The Britons were led by C.W. Kent, a rower who had captained Brasenose College, Oxford, and the Molesey club. They stained their limbs

E.T. Reed, prehistoric Henley Regatta (1894), as reprinted in E.T. Reed, *Prehistoric Peeps* (London: Bradbury and Agnew, 1896)

and faces red and blue and took to the river on a large raft adorned with a tepee and blazing fire, around which they danced primitively.[58] A punt filled with similarly prehistoric 'Gauls' was then propelled into the current. A fight ensued in which the invaders along with an interloping police constable were tossed into the river by the Britons, who then battled a water dinosaur.[59] The event preciously suggested that rowing regattas had existed for ever, which in turn conveyed a humorously nationalist message about how ancient, middle-class, games-playing Britons had repulsed Continental invaders. At the same time, the programme's explicit appeal to the 'gallery gods', those who watched music hall performances from the farthest, cheapest and most raucous balcony, showed the proximity in the public mind between Reed's drawings and the dominant form of mass entertainment.

Similar associations can be seen with cricket, the game most tied to public schools, gentlemanly ideals and empire. Reed satirised it in a July 1894 cartoon, set at Stonehenge, in which he reimagined the trilithons as a prehistoric wicket.[60] The image sent up the public school mythology that the game was pure, manly and indisputably British. This sentiment was expressed in 1897 by C.B. Fry, one of the greatest cricketers of all time, who lauded the game's 'simplicity and strength, freedom and enthusiasm, such as prevailed before things in general became quite as complex and conventional', concluding elegiacally, 'one gets from cricket a dim glimpse of the youth of the world'.[61] Smug self-satisfaction about games made them ideal targets. So it is not surprising that a prehistoric sketch, devised in part by Reed, was staged at the 1898 Oxford–Cambridge swimming races, which were held at a gentlemen's club in Mayfair. The event concluded with a scene in which swimmers clad as peeps fought a floating electric-eyed dinosaur that was worked by two people concealed in its body.[62]

Reed's fancifully archaic versions of Victorian ideals, traditions and rituals set in desolate landscapes resonated especially strongly in imperial outposts, whose inhabitants, as David Cannadine has written, were preoccupied with 'the domestication of the exotic – the comprehending and the reordering of the foreign in parallel, analogous, equivalent, resemblant terms' to the metropolis.[63] Throughout the empire, beplumed and bemedalled administrators presided over Westminster-style parliaments and headed colonial societies that embodied a serious and highly symbolic quest to implant British ideals throughout the world. These explicitly aped upper middle-class Britain with its finely calibrated hierarchy of schools, regiments, clubs and social activities, even if many colonial buildings had corrugated iron roofs and cricket was played on sere and uneven wickets. Magazines like *Punch* were crucial to such societies. The London edition was widely read throughout the empire, while imitative versions were established in many colonies and dominions. Subscribing to such publications was a means of asserting a 'British' identity in the face of emerging colonial ones.[64]

The prevalence of *Punch* throughout the empire meant that colonists recognised parallels between Reed's archaic Britain and their daily experiences. Reed's cartoons gave colonists a vocabulary with which to describe their world that they knew would be understood at home. They invoked the peeps to voice indignation over metropolitan ignorance about conditions in the peripheries, such as the reaction to an 1894 report on Indian sanitary conditions that was criticised in the *British Medical Journal* by a member of the Raj for being 'as comparable with present experience and progress as the "prehistoric peeps" now being furnished for everybody's amusement in *Punch* are with the existing condition of things in Great Britain'.[65] But whimsical self-deprecation was the most common tone in which the peeps were invoked in the empire. Images could be transposed completely, like the revue that was staged in May 1895 to welcome a new chancellor to the University of Sydney, a collegiate institution modelled on an Oxbridge ideal. This included a spoof lecture by a member of the geology faculty which was illustrated with lantern slides of Reed's drawings.[66] Five years later, a British officer serving in South Africa wrote a letter, subsequently published around the globe, in which he described his unit as 'a collection of "prehistoric peeps"'.[67] It was his way of evoking images of the bedraggled and unshaven men as they chased Boer commandos over a veldt that recalled Reed's ancient landscapes. The term also suggested that while the soldiers may have appeared prehistoric, their British ideals were superior to those of their enemies. At almost the same time, a now-lost sketch of a South African cricket match was described in the *St. James's Gazette* as:

> a strange scene, resembling nothing so much as one of Mr Reed's prehistoric peeps. Two weird figures, who are intended to represent the South African Constabulary, face one another on a hillside, which a Lord's ground-man might have envisioned in a severe attack of nightmare. One brandishes a tree branch and the other prepares to swing at him a wooden block chopped into some resemblance of rotundity, while around other strange figures wait with eyes anxiously fixed on the pair.[68]

The description was reprinted in Australia and New Zealand within a month, once again demonstrating how such jokey mental images reinforced a sense of British superiority.[69]

Reed's cartoons were equally brought to life in the colonies through events like the polo match that was staged by the garrison at Quetta, on India's turbulent north-west frontier, in August 1895 (see Figure 10). Military tournaments and gymkhanas were important social events for the British residents of such towns. Men loitered in the marquees and pavilions in their sporting, school, college and club caps, blazers and ties, replicating the spirit of similar events back home. At this match, players dressed like peeps rode mules, camels, donkeys and bullocks that had been painted and adorned to resemble fanciful prehistoric beasts, and used everything from tree roots to camels' skulls as polo sticks. Spectators could consult a giant programme that consisted of pictograms drawn on a stretched sheepskin.[70]

Superficially, colonial officials made references to the peeps to satirise themselves and to project a reassuringly 'British' identity in areas that were very remote from the imperial metropolis. They also saw that their societies and institutions physically resembled the ancient Britain depicted by Reed. At the same time, such events subtly legitimised imperial ambitions by imposing this entirely fanciful British prehistory on areas that had tremendously ancient and complex indigenous societies. A 1901 cartoon published

102   Inventing the cave man

Prehistoric polo, *Graphic* (7 September 1895).

in a Dundee newspaper shows how easily Reed's satiric detachment could become crudely racist, especially when it was invoked outside the British Isles. The drawing accompanied an excerpt from an August 1901 parliamentary report about the East Africa Protectorate, a territory that roughly corresponds to modern Kenya and that Whitehall had recently taken over from the failing British East Africa Company. The author, Sir Charles Eliot, described the railway, which was being constructed from the coast to the fertile interior in the hopes of stimulating agricultural development, in these terms; 'it is not a humorous exaggeration to say that the wayside landscapes remind the traveller of the series of pictures in *Punch*, called prehistoric peeps'.[71] The passage, which distilled a long, formal report into a single, immediately understandable allusion, was reprinted from Toronto and Arizona to Adelaide. The *Dundee Evening Post* accompanied its short acknowledgement of Eliot's report with a faux-peeps drawing depicting the platform at Nairobi's railway station, a reasonable enough editorial decision given Eliot's prose and the centrality of trains in the colony's future (see Figure 11). But the anonymous cartoonist lacked Reed's skill. The result was a cluttered and chaotic scene of elephants charging an incoming locomotive and smashing the booking office, while a huge snake slithers across the railway line. The passengers are thick-lipped, grass-skirted, spear-carrying Africans.[72] Neither the frenzied action nor the racist caricatures bore any relation to Eliot's report, but they exposed the implicit message in Reed's cartoons that equated British ideals alone with civilisation.

Similar depictions of Africans had appeared in *Punch* over the years, including, as we have seen, in Reed's 'Prehistoric Fashoda' cartoon. But

A racist view of prehistory, *Dundee Evening Post* (21 August 1901).   11

Eliot's written descriptions and those of the South African War troops show that British colonials cast themselves in the role of peeps: the middle-class, Anglo-Saxon inhabitants of a primitive land. Many of the British traditions and institutions that they implanted in these colonies resembled the ones that Reed had depicted. Buildings and infrastructure may have been more ramshackle than those at home, while the wild and rugged geography bore little resemblance to the tame, genteel Home Counties idyll. Colonists, administrators and troops found that Reed's drawings provided comforting emotional and psychological ties to the 'mother country'. The images also suggested that the British had a right to push aside indigenous peoples and recreate a society whose institutions, so Reed's cartoons implied, were equally immemorial.

The movement, drama and farce in Reed's drawings and their stage-like settings made the theatre another self-evident place in which to recreate them. This may first have happened in 1894, when Cambridge undergraduates recreated individual cartoons as statically posed *tableaux vivants* in smokers and revues. In the following year they clad native characters as peeps for *Iphigenia of Tarsus*, the 1895 annual Greek play. A theatrical production performed entirely in an ancient Classical language embodied Victorian educational ideals, but inserting the peeps into the cast was a whimsical way of subverting this didactic seriousness.[73]

A more publicly accessible and rollicking appropriation was *Boadicea Unearthed*, a musical comedy staged by professional companies at the start of 1895 in Hull and Kilburn and soon thereafter by amateur dramatic societies. It married the craze for the peeps with the excitement that had surrounded the recent discovery of what was believed to be the burial mound of Queen Boudicca, who had led a Celtic revolt against the Romans, only to kill herself when it faltered. She was one of the greatest heroes of the Victorian age, thanks in part to a laudatory poem by Tennyson and an as-yet unfinished monumental statue, commissioned by Prince Albert, that now adorns Westminster Pier. The supposed discovery of concrete evidence of her existence propelled her from the realm of myth into an exemplar of Britain's prehistoric past. The play opened with the excavation of Boadicea's tomb by an antiquary, assisted by a masher, or preening and affected young man, and 'a bespectacled pseudo-scientific Modern Woman' such as the universities were producing, much to satirists' delight. The trio were startled to find that Boadicea shared her grave with a cave man, described by one reviewer as having been 'conveyed from the pages of *Punch*'.[74] The placing of both figures in the tomb explicitly equated Britain's heroic and archaeologically sound past with Reed's imagined one. Boadicea and the peep came to life and fell in love with their polar opposites: the warrior queen courted the young man, while the peep was attracted to the intellectual woman. Mismatched love was not easy and, after singing a duet that spoofed current songs, the unearthed pair admitted their unhappiness, suggesting the Victorian age was not worthy of Britain's storied past. They were then reburied, unconsoled and broken-hearted.[75]

The most distinguished theatrical evocation of the peeps occurred in *The Rose of Persia*, a Savoy opera by Basil Hood and Sir Arthur Sullivan that was premiered in November 1899. It was set in the house of a wealthy Middle Eastern merchant with a plot that featured mistaken identity, dancing girls and a drug-induced dream. The libretto of this much-needed success for the D'Oyly Carte company included the lines:

> Let a satirist enumerate a catalogue of crimes
> Though he label them the outcomes of our shallow modern times;
> Yet a Persian *Punch*'s pencil, in a prehistoric peep,
> Would show us human nature just as shallow – or as deep.[76]

Critics drew readers' attention to these topical references, for which the company was famous, knowing they would serve as a calling card for the production.[77]

Reed's drawings also inspired writers to explore the fictional possibilities of placing modern humans in prehistory. The first to do so appears to have been Rudyard Kipling, the writer most closely identified with Britain's imperial ideals. In 1894 he published a thirteen-stanza poem, entitled 'The story of Ung', about a Cro-Magnon artist who resents his tribe's inability to appreciate his paintings.[78] A year later Kipling wrote 'In the Neolithic Age', in which the work of an ancient artist is discovered by a modern-day researcher and becomes the subject of verses by a 'minor poet certified by Traill'. It is a sly reference to Henry Duff Traill, who was both a prominent critic and the author of one of the first literary evocations of British prehistory.[79] Kipling was jokily exploring art's place in society by projecting the conflict into prehistory. The discovery of bone carvings and cave paintings over the previous decades had shown Cro-Magnon's advanced artistic skills.[80] So Kipling was using the same distancing device as Reed did in his cartoon. It is also interesting to note that the works belong to the period in which he produced *The Jungle Book* and *The Second Jungle Book*.

While these are not exactly prehistoric stories, there are obvious analogies between the peeps and Mowgli, the British child raised by animals in the Indian jungle. Both men played broadly on the theme that modern British sensibilities were innate, immutable and superior.

Despite Kipling's efforts, poetry would never be prehistory's literary home. The foremost prose writer to take up the subject was undoubtedly H.G. Wells, the pioneer of science fiction and one-time student of Huxley. He was as keenly attuned to the popular imagination as Kipling. He also belonged to the literary circle around Edward Clodd, a wealthy banker who had written two best-selling explanations of evolution. Clodd and Wells were fascinated by 'primitive' societies and the role of culture in civilisation.[81] Wells most famously broadcast these ideas in his 1895 novel *The Time Machine*. However, he explored the fictional possibilities of the ancient past more concretely two years later in 'A story of the Stone Age', which was serialised in the *Idler* magazine. It follows the lovers Ugh-Lomi and Eudena, whose tribe inhabits a densely forested primeval Surrey. Ugh-Lomi's prowess as a hunter is greatly increased when he invents a crude axe by lashing a flint rock to a stick. This emboldens him to fight a cave-bear in an inconclusive contest, at the end of which Wells shifts into an uncharacteristically comic tone, reflecting the dominant public perception of prehistory. After the contest, the stunned cave-bear admits to his mate:

> 'I never was so startled in my life,' he said at last. 'They are the most extraordinary beasts. Attacking me!'
> 'I don't like them,' said the she-bear, out of the darkness behind.
> 'A feebler sort of beast I never saw. I can't think what the world is coming to. Scraggy, weedy legs … Wonder how they keep warm in winter?'
> 'Very likely they don't,' said the she-bear.
> 'I suppose it's a sort of monkey gone wrong.'
> 'It's a change,' said the she-bear.[82]

This short exchange evokes images of Reed's gangly and awkward characters, which had monopolised the public's idea of cave men, while also sending up the notion that equated humanity's evolution from the apes with progress.

Over the following years, less accomplished writers also explored Britain's imagined prehistory in books bereft of humour but filled with didactic purpose. Among them was Stanley Waterloo, one of the first American authors to enjoy immense success in Britain. In 1897 he published *The Story of Ab*, a novel about love, rivalry, monsters and the invention of the bow and arrow, set in a prehistoric Thames Valley. Waterloo constantly described his creatures with mocking references to Victorian middle-class pretensions, working roughly the same furrow as Reed. In the novel's final pages, Ab's tribe intermarries with Celts, Gauls and Saxons to forge the British nation.[83] Waterloo's clear intention for British readers to identify with his creatures can also be detected in John Preston True's 1899 book for young people, *The Iron Star*. It is a serious story about a prehistoric brother and sister named Umpl and Sptz, whose semi-pronounceable names acknowledge the pre-linguistic state of their tribe. Their outcast family discovers fire when a flaming meteorite – the iron star of the title – crashes to earth. They smelt metal and are propelled into the Iron Age, while Umpl becomes a great hunter and war leader. The book was a thinly veiled teleological lesson in British history which followed the protagonists' descendants through the Viking invasions and Magna Carta to the reign of Queen Victoria.[84]

Wells, Waterloo and True were inspired at least in part by the interest in British prehistory that Reed had so greatly popularised. However, the public conflation of Reed's comically imagined past and more rigorous knowledge offended at least one prominent populariser of science. The Cambridge-educated Reverend Henry Neville Hutchinson had abandoned the rectory to write science texts for the mass market. Books by Hutchinson like *Extinct Monsters*, an influential 1892 explanation of dinosaurs, were noted for their accurate illustrations. Hutchinson was especially incensed by the way in which Reed intermingled humans and dinosaurs and the fanciful licence he took in drawing the latter. The peeps were especially galling to Hutchinson because he and Reed aimed their works at much the same middle-class audiences.[85] So Hutchinson approached Reed soon after the first peeps drawings appeared, suggesting that his own works might provide tips about prehistoric anatomy. Reed freely admitted reading *Extinct Monsters*, though he was infuriated by Hutchinson's repeated claims that the peeps were derived entirely from his books.[86]

In 1899, Hutchinson took Reed on directly with a volume of scientifically accurate prehistoric cartoons. He was no artist, so he hired the illustrators John Hassall – who is best known for a 1908 poster showing a fisherman skipping down a beach along with the words 'Skegness is so bracing' – and Frederick Burridge to produce twenty drawings. The resulting book, *Primeval Scenes: Being Some Comic Aspects of Life in*

Prehistoric Times, appeared that Christmas. Advertisements proclaimed that 'the author has taken great pains to show the public that "prehistoric peeps" may be amusing, and at the same time drawn strictly in accordance with scientific discovery', signalling Hutchinson's sober intent and grumpy tone.[87] Hassall and Burridge had striven to replicate Reed's delight in human folly, but they lacked his satirical intent and sense of whimsy and produced only cluttered scenes that intermixed prehistoric humans with temporally appropriate and anatomically and evolutionarily correct mammals. Everyone could see that the images were derived from the peeps, to which Hutchinson's volume was inevitably and negatively compared.[88]

For all its tetchy pedantry, Hutchinson's reaction to the peeps can be taken as an analogy for the relationship between the late Victorian Church and evolution. The equanimity with which this ordained priest greeted Reed's version of prehistory would have been very unlikely in the 1860s. However, as a scientist Hutchinson deplored the cavalier mixture of dinosaurs, extinct mammals and Cro-Magnons. Evidence-based and accurate knowledge of the prehistoric past had become Hutchinson's inflexible orthodoxy.

It is impossible to dismiss Reed's very significant influence on subsequent depictions of prehistory. The peeps cartoons dominated popular representations of the prehistoric past around the world almost from their first appearance. They were created in the imperial metropolis and transmitted through a mature mass culture that crossed artistic forms, social and educational hierarchies and physical distance. Reed made prehistory a fundamentally British era that minimised evolutionary concepts in favour of an archaic version of the contemporary world where humans, ancient mammals and dinosaurs were jumbled together. Reed and his many imitators either simply erased non-Europeans from prehistory or relegated them to subordinate roles that buttressed late Victorian racist beliefs and supported the implantation of British ideals and institutions in distant lands. The cartoons' popularity reflected the way in which they reinforced popular assumptions of British, and by extension Anglo-Saxon, superiority.

In what would become a recurring commentary about the effect of Reed's cartoons, a review of an exhibition by members of the Royal Institute of Painters in Water Colours in 1901 pointed out that they made it difficult to see more sober attempts to depict prehistory as anything but funny.[89] The pudding was certainly being overegged for effect, but such comments showed how influential the comic view of prehistory had become by the turn of the century. Scientifically correct recreations of the prehistoric past were consequently relegated to museums and textbooks, where they have remained for the most part. As the next chapter shows, interest in the peeps continued into the twentieth century, while an equally popular music hall evocation of the drawings laid the foundation for virtually all subsequent representations of cave men in global mass culture.

Notes

1 Pat Shipman, *The Man who Found the Missing Link* (London: Simon & Schuster, 2001), passim; 'Antediluvian remains in France', *John Bull* (23 March 1894), p. 190; W.K. Marischal, 'The missing link at last', *English Illustrated Magazine* (March 1897), pp. 653–7.
2 The quotation is from Adrian Desmond, *Huxley: From Devil's Disciple to Evolution's High Priest* (Reading: Addison Wesley, 1997), p. 225. See also Sir Arthur Keith, *An Autobiography* (New York: Philosophical Society, 1950), pp. 172–4 and 193; and Peter C. Kjærgaard, '"Hurrah for the missing link!": a history of apes, ancestors and a crucial piece of evidence', *Notes and Records: The Royal Society Journal of the History of Science*, 65:1 (2011), pp. 91–2.
3 'Surrey Theatre', *Stage* (22 March 1894), p. 10; 'The missing link', *Stage* (29 March 1894), p. 12; 'Barnum and Bailey', *Penny Illustrated Paper* (17 December 1898), p. 395; 'Barnum's fre – –, pardon, prodigies awheel', *Cycling* (28 January 1899), p. 45.
4 'Discovery of the missing link', *Moonshine* (19 January 1895), p. 34.
5 ''Enery and the missing link', *Judy* (23 January 1895), p. 48.
6 'Amusements', *Burnley Gazette* (15 February 1890), p. 1; 'The People's Palace', *Dundee Advertiser* (1 October 1891), p. 1; 'Theatre Royal', *Belfast Newsletter* (9 April 1895), p. 5; 'Amusements in Leeds', *Era* (10 August 1895), p. 15; and 'Gilbert's modern circus', *Nottingham Evening Post* (15 November 1899), p. 1.
7 'Theatrical gossip', *Era* (12 November 1892), p. 10; 'Cycle parade at the Hartlepools', *Northern Echo* (24 July 1893), p. 4; 'Cycle parade at Darlington', *Northern Echo* (29 July 1893), p. 4.
8 Samuel Page Widnall, *A Mystery of Sixty Centuries, or a Modern St. George and the Dragon* (Grantchester: S.P. Widnall, 1889); and Bruce Dickins, 'Samuel Page Widnall and his press at Grantchester, 1871–1892', *Transactions of the Cambridge Bibliographical Society*, 2.5 (1958), pp. 366–72.
9 H.D. Traill, 'A day with primeval man', *Universal Review* (March 1889), pp. 304–16; Pierre Boitard, *Paris avant les hommes* (Paris: Passard, 1861).
10 Louis Figuier, *Primitive Man* (London: Chapman and Hall, 1870).
11 *Vénus et Caïn: Figures de la préhistoire, 1830–1930* (Bordeaux: Éditions de la réunion des musées nationaux, 2003), passim; Stephanie Moser, *Ancestral Images: The Iconography of Human Origins* (Ithaca: Cornell University Press, 1998), pp. 1–6.

12 Richard Noakes, '*Punch* and mid-Victorian comic journalism', in Geoffrey Cantor et al., *Science in the Nineteenth Century Periodical: Reading the Magazine of Nature* (Cambridge: Cambridge University Press, 2004), p. 93; Moser, *Ancestral Images*, pp. 3–18.
13 'Prehistoric peeps', *Punch's Almanack for 1894* (1893), unpaginated. *Punch's Almanack*s were published in December, in time for Christmas sales. So the *Almanack* for 1894 actually appeared in the last weeks of 1893.
14 'Prehistoric peeps', *Punch* (23 December 1893), p. 292.
15 See for instance 'Peeps at Paris', *Punch* (16 March 1867), p. 105.
16 Roy Compton, 'A chat with Mr E.T. Reed', *Idler* (May 1896), pp. 505–6.
17 'A warning to enthusiasts', *Punch* (6 July 1889), p. 5. The quotation is from 'A chat with Mr Harry Furniss's successor', *Launceston Examiner* (Australia) (6 June 1894), p. 7.
18 'Evolutionary assimilation', *Punch* (12 July 1890), p. 21.
19 'Professor March's primeval troupe', *Punch* (13 September 1890), p. 124. See also 'British Association', *Leeds Mercury* (5 September 1890), p. 6.
20 'A comparatively quiet night in the primeval parliament', *Punch* (14 April 1894), p. 178.
21 The quotation is from Compton, 'A chat with Mr E.T. Reed', p. 508. See also untitled, *Launceston Examiner* (Australia) (6 May 1905), p. 9.
22 The quotation is from 'A chat with Mr Harry Furniss's successor', p. 7; see also Raymond Blathwayt, 'How I do my *Punch* pictures: a talk with Mr E.T. Reed', *Windsor Magazine* (7 December 1897), p. 450; 'A maker of mirth', *Launceston Examiner* (Australia) (12 April 1898), p. 5; and E.T. Reed, 'Prehistoric peeps', *The Times* (26 December 1899), p. 8.
23 'Sam Sly's comic picture gallery: a prehistoric romance', *Boy's Comic Journal* (7 July 1888), p. 304. For a description of the magazine see James Chapman, *British Comics: A Cultural History* (London: Reaktion Books, 2011), p. 18.
24 'Gorilla warfare under the protection of the American flag', *Puck* (19 March 1884), p. 1; 'A prehistoric romance', *Puck* (20 August 1890), p. A27.
25 The quotation is from 'Christmas numbers', *Morning Post* (13 December 1893), p. 2. See also 'News in brief', *South Wales Daily News* (29 November 1893), p. 4; 'Echoes of the week', *Leeds Mercury* (30 December 1893), p. 12; 'Summary of news', *Sheffield Daily Telegraph* (2 February 1894), p. 4; '*Punch* and the Commons', *Hartlepool Mail* (12 April 1894), p. 3; and 'Lika Joko', *Star* (Guernsey) (3 November 1894), p. 2.
26 Peter Bailey, 'Conspiracies of meaning: music-hall and the knowingness of popular culture', *Past & Present*, 144 (1994), pp. 138–70.
27 'Prehistoric pantomime', *Punch Almanack for 1895 (1894)*, unpaginated; and 'Prehistoric waits', *Punch* (26 December 1896), p. 310.
28 E.T. Reed, 'A night lecture on evolution', *Punch* (23 June 1894), p. 298.
29 Edward Caudill, 'The press and tails of Darwin: Victorian satire of evolution', *Journalism History*, 20:3 (1994), p. 110.

30 See for instance 'Ask a white man!', *Punch* (14 June 1890), p. 280; and 'Collapse of corner men', *Punch* (13 September 1890), p. 131.
31 See for instance the photograph of Reed in the National Portrait Gallery collection, NPG x44937, http://www.npg.org.uk/collections.php (last accessed 6 October 2016).
32 'Prehistoric Scotland de-picted', *Punch's Almanack for 1900 (1899)*, unpaginated; 'A primeval yacht race', *Punch* (2 October 1901), p. 249; 'Ping-pong in the stone age', *Punch* (4 September 1901), p. 177; 'Unity and unanimity', *Punch* (11 December 1901), p. 122.
33 'Survivors of the (glad-)stone age', *Punch* (1 May 1901), p. 333; 'Sir H. Campbell-Bannerman on the war and its cost', *Derby Daily Telegraph* (25 April 1901), p. 2.
34 Henry J. Miller, 'John Leech and the shaping of the Victorian cartoon: the context of respectability', *Victorian Periodicals Review*, 42:3 (2009), pp. 276–7.
35 E.T. Reed, *Prehistoric Peeps* (London: Bradbury and Agnew, 1896).
36 The quotations are from E.T. Reed, *Mr Punch's Animal Land* (London: Bradbury and Agnew, 1898), unpaginated; see also untitled, *Athenaeum* (26 November 1898), p. 736.
37 E.T. Reed, *Punch's Holiday Book* (London: Bradbury and Agnew, 1901), p. 6.
38 'Interesting bas-reliefs', *Nottingham Evening Post* (10 March 1899), p. 2; G.C. Williamson, 'François Lutiger and his silver work', *Artist* (May 1899), pp. 72–5.
39 See for instance 'An artistic causerie', *Graphic* (11 March 1899), p. 306; 'World of art', *Glasgow Herald* (29 June 1899), p. 7; 'Wales in parliament', *South Wales Daily News* (30 June 1899), p. 5; and 'London week by week', *Leeds Times* (8 July 1899), p. 3.
40 *Punch* archives, British Library, Add. MS 88937/1/2, letter book, Philip L. Agnew to Reed, 20 September 1894.
41 See for instance William A. Dutt, 'Ung the cave dweller', *Pall Mall Gazette* (17 November 1898), pp. 1–2.
42 See for instance 'The boat-race in ancient Egypt', *Punch* (3 April 1897), p. 167; 'Unrecorded history', *Punch* (31 October 1896), p. 214; and 'The tablets of azit-tigleth-miphansi, the scribe', *Punch* (29 August 1900), p. 157.
43 For Reed's income and addresses see *Punch* archives, British Library, various letter books, 1893–1904. See also Michael Heller, 'Work, income and stability: the late Victorian and Edwardian London male clerk revisited', *Business History*, 50:3 (2008), pp. 253–71; and Helen C. Long, *The Edwardian House: The Middle Class Home in Britain, 1880–1914* (Manchester: Manchester University Press, 1993), pp. 8–9.
44 Frederick Dolman, 'Our graphic humorists: their funniest pictures as chosen by themselves', *Strand Magazine* (January 1902), p. 80.
45 For 'Ted' see *Punch* archives, British Library, Add. MS 88937/1/5, letter book, Philip L. Agnew to Reed, 12 December 1898.
46 See for instance '*The Windsor Magazine*', *Warwick Argus* (Queensland) (26 April 1898), p. 2; 'Mr E.T. Reed of *Punch*', *Dundee Evening Telegraph* (11 September

1900), p. 6; E.T. Reed, 'Humour in black and white: a sketch', *Magazine of Art* (December 1900), pp. 117–22; 'Sheffield notes and jottings', *Sheffield Evening Telegraph* (22 October 1901), p. 5; 'Armistead lectures: a caricaturist and his work', *Dundee Evening Telegraph* (23 November 1901), p. 4; and 'A *Punch* caricaturist on his art', *Perth Daily News* (Australia) (2 January 1902), p. 5.

47 See for instance 'Extract of zoology', *Pall Mall Gazette* (26 April 1897), p. 9; 'Interesting lectures at Blackburn Technical School', *Blackburn Standard* (24 December 1898), p. 5; 'Excavations at Largo', *Dundee Evening Telegraph* (14 May 1901), p. 3; and 'A prehistoric peep', *Gloucester Citizen* (1 January 1906), p. 4.

48 'The man about town', *South Wales Echo* (6 July 1894), p. 2; 'Lincoln board of guardians', *Stamford Mercury* (7 December 1894), p. 6; 'The game and the gale', *Yorkshire Evening Post* (24 December 1894), p. 3; 'A prehistoric peep', *Sunderland Daily Echo and Shipping Gazette* (9 January 1895), p. 2; 'Chester's new baths', *Cheshire Observer* (30 June 1900), p. 8; 'The pantomimes', *Manchester Courier and Lancashire General Advertiser* (28 December 1901), p. 5.

49 'Clyde dock scheme', *Glasgow Herald* (11 July 1899), p. 7; 'Sheffield city council', *Sheffield Daily Telegraph* (24 October 1901), p. 7.

50 G.E.H Barrett-Hamilton and H.O. Jones, 'A visit to Karaginski Island, Kamchatka', *Geographical Journal*, 12:3 (1898), p. 298.

51 'Lunch and prehistoric peeps', *Pall Mall Gazette* (31 July 1899), p. 8.

52 See for instance 'The 'life of *Punch*', *Littell's Living Age* (11 January 1896), p. 93; W.D. Stevens, 'Prehistoric football in the jungle', *Century Illustrated Magazine* (May 1897), p. 160; untitled, *Life* (27 July 1899), p. 72; 'Literary notes', *Albany Law Journal* (January 1903), p. 28; 'Prehistoric peeps', *Frank Leslie's Popular Monthly* (January 1895), p. 124; and Compton, 'A chat with Mr E.T. Reed', p. 505.

53 'Prehistoric peeps', *Salt Lake Herald* (Utah) (27 January 1895), p. 7.

54 'A literary chemist', *News Herald* (Ohio) (21 February 1895), p. 2; 'A literary chemist', *Hawaiian Star* (30 May 1895), unpaginated.

55 'Collegiate school swimming sports', *Wanganui Chronicle* (4 March 1895), p. 2; 'Exhibition of photographic slides', *Nelson Evening Mail* (7 October 1899), p. 2.

56 'Sporting notes and news', *Pall Mall Gazette* (21 August 1894), p. 9; 'Prehistoric peeps', *Punch* (7 July 1894), p. 10.

57 'Up the river', *London Daily News* (1 September 1894), p. 6.

58 'Rowing', *Baily's Magazine of Sports and Pastimes* (July 1893), pp. 47–8.

59 'The world of sportswomen', *Hearth and Home* (30 August 1894), p. 556.

60 E.T. Reed, 'Prehistoric peeps', *Punch* (21 July 1894), p. 34.

61 C.B. Fry, 'Cricketers I have met', *Windsor Magazine* (June 1897), p. 152.

62 'Bath club', *Morning Post* (7 May 1898), p. 8; 'Letter from our London correspondent', *Newcastle Journal* (15 July 1898), p. 5.

63 David Cannadine, *Ornamentalism: How the British Saw their Empire* (Oxford: Oxford University Press, 2001), p. xix.

64 Richard Scully, 'A comic empire: the global expansion of *Punch* as a model publication, 1841–1936', *International Journal of Comic Art*, 15:2 (2013), pp. 15–18.
65 G. Hutchinson, 'Hindu pilgrimages in India', *British Medical Journal* (20 October 1894), p. 907.
66 'University conversazione', *Sydney Evening News* (30 May 1895), p. 4. See also 'The science and art of coaching', *Queenslander* (27 July 1895), p. 158; and 'Prehistoric peeps', *Albury Banner and Wodonga Express* (23 August 1901), p. 32.
67 See for instance 'Pursuit of De Wet', *The Times* (22 November 1900), p. 10; and 'In pursuit of De Wet', *Montreal Gazette* (6 December 1900), p. 10.
68 'Imperial sport', *St. James's Gazette* (13 September 1901), pp. 3–4.
69 'Imperial sport', *Sydney Morning Herald* (30 October 1901), p. 10; 'Imperial sport', *Oamaru Mail* (9 November 1901), p. 4.
70 'Prehistoric polo at Quetta', *Graphic* (7 September 1895), p. 296.
71 Sir Charles Eliot, *Report by Her Majesty's Commissioner on the East Africa Protectorate* (London: His Majesty's Stationery Office, 1901), p. 19. The passage was quoted in 'The East Africa protectorate', *Sheffield Evening Telegraph* (19 August 1901), p. 4; 'An African railway', *Toronto Globe* (4 September 1901), p. 6; 'Concerning people', *Register* (Adelaide) (4 October 1901), p. 5; and 'Humours of a blue book', *Arizona Weekly Journal-Miner* (6 November 1901), unpaginated.
72 'Civilisation in central Africa', *Dundee Evening Post* (21 August 1901), p. 6.
73 Smokers and revues are in Raymond Blathwayt, 'How I do my *Punch* pictures: a talk with Mr E.T. Reed', *Windsor Magazine* (7 December 1897), p. 444. The Greek play is in 'The *Iphigenia of Tarsus* at Cambridge', *Athenaeum* (8 December 1894), p. 800.
74 'Amateurs', *Stage* (7 February 1895), p. 18.
75 'Lunchwright Dramatic Society', *Era* (2 February 1895), p. 11; 'New plays of the month', *Era* (9 February 1895), p. 9; Wendy C. Nielsen, *Women Warriors in Romantic Drama* (Wilmington: University of Delaware Press, 2013), p. 147. See also Jim Davis, 'Imperial transgressions: the ideology of Drury Lane pantomime in the late nineteenth century', *New Theatre Quarterly*, 12:46 (1996), p. 152.
76 Basil Hood and Arthur Sullivan, *The Rose of Persia, or the Story-Teller and the Slave* (London: Chappell, 1899), p. 34.
77 'The new Savoy opera', *Whitstable Times and Herne Bay Herald* (9 December 1899), p. 5; 'Advertisements and notices', *Glasgow Herald* (19 November 1898), p. 1.
78 Rudyard Kipling, 'The story of Ung', in *Rudyard Kipling's Verse: Inclusive Edition* (New York: Doubleday, 1918), p. 397.
79 Rudyard Kipling, 'In the Neolithic Age', in *Kipling's Verse*, p. 394.
80 Glyn Daniel and Colin Renfrew, *The Idea of Prehistory* (Edinburgh: Edinburgh University Press, 1988), pp. 51–3. See also 'Pictures painted 25,000 years ago: works by stone-age artists', *Illustrated London News* (10 August 1912), p. 222.

81 Richard Pearson, 'Primitive modernity: H.G. Wells and the prehistoric man of the 1890s', *Yearbook of English Studies*, 37:1 (2007), pp. 58–61.

82 H.G. Wells, 'A story of the Stone Age', in *Tales of Space and Time* (London: Macmillan, 1920), p. 104.

83 Stanley Waterloo, *The Story of Ab* (Chicago: Way and Williams, 1897).

84 John Preston True, *The Iron Star* (London: Gay and Bird, 1899).

85 H.N. Hutchinson, *Extinct Monsters* (London: Chapman and Hall, 1897). Succinct discussions of the book are found in Bernard Lightman, *Victorian Popularizers of Science: Designing Nature for New Audiences* (Chicago: University of Chicago Press, 2007), pp. 450–60; and Stephanie Moser and Clive Gamble, 'Revolutionary images: the iconic vocabulary for representing human antiquity', in Brian Leigh Molyneux (ed.), *The Cultural Life of Images: Visual Representation in Archaeology* (London: Routledge, 1997), p. 202.

86 'Extinct Monsters in *Punch*', *South Wales Daily News* (14 August 1895), p. 6; 'A maker of mirth', p. 5; E.T. Reed, 'Prehistoric peeps', *The Times* (26 December 1899), p. 8.

87 'Publishers' announcements', *Pall Mall Gazette* (15 December 1899), p. 9.

88 H.N. Hutchinson, *Primeval Scenes: Being Some Comic Aspects of Life in Prehistoric Times* (London: Lamley, 1899); 'Primeval scenes', *Morning Post* (30 November 1899), p. 3; 'Primeval scenes for children', *Review of Reviews* (December 1899), p. 626; 'Primeval scenes', *Graphic* (20 January 1900), p. 96.

89 'The Royal Institute of Painters in Water Colours', *Sketch* (27 March 1901), p. 398.

5

He of the auburn locks: George Robey, the Edwardian cave man

In April 1902, an enlarged compendium of E.T. Reed's 'Prehistoric peeps' cartoons was published. These comic scenes of cave men had first appeared in *Punch* almost a decade earlier and had subsequently been reprinted throughout Britain, the empire and the United States along with many imitative images. Among these was the August 1901 cartoon in Cardiff's *Weekly Mail* in which a slight, gangly cave man, obviously none the better for drink, stands unsteadily at the counter of a cave-like public house. A burly policeman entering the premises orders the publican not to serve the peep. The image satirised the chief constable's recent declaration that his men had greatly reduced public drunkenness in the city, by suggesting that such behaviour was an age-old and innocent pastime. The officious constable represents the forces of modern rectitude. Such cartoons and the constant spontaneous references to cave men on stage and in colloquial speech throughout Britain and abroad had cemented Reed's idea that prehistory was a comic, archaic version of the contemporary world.[1]

The view took greater hold on the public imagination at the start of the century, which corresponded almost exactly with the ascension of the first new monarch in six decades and the introduction of technologies – Dreadnought battleships, moving pictures, motor cars and aeroplanes among them – that would revolutionise Western society. At the same time, national and imperial self-confidence were being challenged by the stalemate of the South African War, competition from emerging empires and the supposedly noxious effects of urban life at home. Cassandras proclaimed that the British 'race' – to use the term of the day – was degenerating. The sense of British exceptionalism was also tested by discoveries of proto-human remains on the Continent that pointed to a sophisticated, pan-European ancient past.[2] If Britain seemed slightly less secure, Reed's vision of prehistory reassuringly asserted that the country's ideals and institutions were eternal.

The most important Edwardian evocation of the peeps coincided with Reed's new book, as George Robey premiered a comic sketch entitled 'The prehistoric man' at the Pavilion, one of London's grandest music halls.[3] The roots of Robey's character were immediately clear. In a gushing review, the *Era* exclaimed that Robey had 'struck oil' with this sketch about a man 'who is supposed to have belonged to some period known only to the learned in geology and biology. In make-up the comedian may be said to reproduce the traditional garments of the aborigines of these islands worn at the time when Julius Caesar was quite a boy, and long before Boadicea, queen of the Iceni, fell before the Roman legions.'[4] The plot was a simple love triangle in which Robey played 'He of the auburn locks', a young man vying with a less evolved rival named 'He of the knotted knee' for the affections of a maid called 'She of the ceaseless tongue'. In the climax, Robey clubbed his opponent and tossed his body into the sea, before singing triumphantly that stones were used as currency in prehistoric times, and that keeping the murder out of the newspapers – the tabloid press was only about a decade old – had cost him the equivalent of the Giant's Causeway, the natural wonder consisting of forty thousand basalt stones on the Northern Irish coast.[5]

The plot seems familiar to twenty-first-century readers because it has been enacted in myriad cave man sketches ever since. It was a fanciful weave of idealised love with so-called 'courtship with a club', or the custom of forced marriage then being explored by the influential Finnish anthropologist Edward Westermarck. He argued that all primitive societies pass through a stage in which such courtship rituals prevail. The belief, which anthropologists have long since abandoned, remains in popular culture as the now-clichéd scene in which a cave man drags a love-struck woman away by her hair.[6] On a more mundane level, audiences would have interpreted the sketch's knockabout violence, in which the slight Robey overcame a bigger rival, as yet another suggestion of Britain's eternal supremacy.

Robey and Reed exemplified contented, middle-class Britishness to overlapping audiences. Unlike many music hall artistes, the thirty-two-year-old Robey, whose real name was George Wade, came from a middle-class family and cultivated an image of contented, suburban family life. The reality was more prosaic, though no less bourgeois. Robey had initially emulated his father by working as a clerk with a Birmingham engineering firm, but gave it up to become a professional performer in 1892, building a stable of sketches over the next decade in which he played boffins, petty officials and historical figures, reflecting his own roots in a Midland Pooterdom that would probably have been recognisable to a large segment of his audience. Robey's sketches turned on punning language, malapropisms and bowdlerised knowledge of elite ideas and customs. No matter what the costume, he sported a pair of heavy grease-paint eyebrows and asserted a dignified modesty by giving the audience a faintly shocked command to 'desist' when they tittered. His on-stage sobriety epitomised the conservative, middle-class aspirations of the men who controlled variety entertainment. By 1902, Robey was one of Britain's most popular comedians, and was widely called 'the prime minister of mirth', a nickname that evoked patriotism and respectability in equal measure.[7]

In other words, the generalised scientific and historical knowledge, middle-class satire and knockabout comedy that were combined in Reed's prehistoric cartoons were natural fits for Robey. As the *Era* proclaimed shortly after the sketch's debut, 'the satire at the expense of modernity is often pungent'.[8] Despite Robey's fairly obvious inspiration, he told an evasive story about stumbling onto the character while indulging in a suitably fashionable and expensive hobby, writing that the 'idea came by chance' and adding, 'I am a photographer and develop prints myself. One morning for a lark I stripped to the waist, put on a wig and whiskers and got someone to snap me. When I developed it and showed it to a friend who happened to be a song-writer he looked at it and said "My goodness, you look like a prehistoric man. I'll write a song on that!"'[9] By the time Robey first performed the sketch, his costume consisted of spiky red hair and beard, a shaggy animal hide draped across one shoulder and an axe fashioned from a large rock lashed to a stick. Only Robey's trademark eyebrows betrayed his true identity. It was clear to audiences, as Robey must have anticipated, that this cave man was, as the *Era* put it, 'founded on Mr Reed's wonderfully clever sketches in *Punch*'.[10] Robey nonetheless inflected the character with his own comic sensibilities by silencing those who shouted out comments from the auditorium – a common event in music hall – with a stern 'they had manners in those days' and 'they had pure minds in those days'.[11] It is satisfying, if speculative, to imagine that Robey employed the first retort when someone in the audience compared him to a peep, and the second with those offering more salacious advice to his courting cave man. Londoners loved the sketch, which Robey performed from Camberwell and Brixton in the south, eastwards to Stratford and westward to the lavish halls around Leicester Square over the summer, capturing a wide socio-economic swathe of the capital's inhabitants.[12]

Reports about Robey's sketch made it much more than a London sensation. He played it during the summer in provincial centres like Liverpool, Leicester, Cardiff and Nottingham, where a reviewer restated the self-evident point that Robey was 'elaborating an idea which enabled Mr E.T. Reed to furnish the readers of the leading comic journal with a wealth of pictorial humour'.[13] Robey made the first of several recordings of the sketch that summer, tapping into the growing popularity of music in the home. Not everyone could afford to spend more than £2 on a gramophone or 2s on a recording, but at the turn of the century music hall artistes realised that the new industry was a way of making extra money and expanding their fame.[14] Today, it is fairly easy to hear Robey singing 'The prehistoric man' online and in commercial compilations of music hall favourites, though these only hint at what Edwardians experienced. Music hall relied fundamentally on live performances, as artistes subtly altered the pacing of sketches and songs after gauging an audience's mood, and interjected topical references and direct addresses to those in the theatre. Nonetheless, live and recorded versions of Robey's sketch were so ubiquitous by mid-summer that the magazine *Stage* quipped that it was becoming 'just a wee bit too prehistoric'.[15] The judgement was not universal, as Robey performed the sketch until year's end and revived it constantly thereafter because, so he claimed, it was 'the first song that brought me any money worth speaking of' thanks to his salary, recordings and the thousands of photographic postcards he sold of himself in costume, posing before a backdrop depicting a barren wasteland (see Figure 1).[16]

The sketch also reinforced the delineation between depictions of cave men derived from Reed's cartoons and the artistes who continued performing as acrobatic 'man-monkeys' in the style that had been pioneered in the 1860s. Such acts were based on out-of-date ideas about infinitely ancient hominids and evolution, though the frequency of advertisements for performances in places like Hull, Belfast, Ilkeston, Stockport, Hartlepool and Ashton-under-Lyne demonstrated both the continuing pull of nostalgic, unfashionable entertainment and its banishment to the peripheries of the industry.[17] Some of these acts tried to co-opt Robey's fame directly. For instance, Consul, a chimpanzee clad in gentleman's attire – shades of *Punch*'s 'Mr Gorilla' – was billed as 'The prehistoric man' for a performance in which it carried out mundane domestic activities at the London Hippodrome in 1903, while the Abadaroffs, Manchester acrobats, toured Britain for years with advertisements that compared them to Robey or touted them as 'Prehistoric peeps'.[18]

Robey alone could not meet the demand for cave man sketches in the wake of his debut. As one of the era's most popular artistes, he appeared in only the largest halls and cities, leaving ample room for imitations and evocations throughout the country. Within weeks of his first appearance, a Celtic opera composed and performed by employees of the London & Westminster Bank was staged in London at St George's Hall, Langham Place. This decidedly amateur production included scenes in which ancient Britons were clad in prehistoric costumes.[19] A far more blatant response to Robey was heard in mid-summer when Marie Lloyd, another of the era's greatest stars and the embodiment of Cockney cheek, premiered a song entitled 'The prehistoric woman' at the Tivoli Music Hall in the Strand. Lloyd was appearing in a 'revue', a theatrical concept recently imported from Paris that, as a Manchester newspaper explained, consisted of 'a series of scenes, [in which] topical events are mirthfully parodied and well-known personages good-humouredly caricatured'.[20] Lloyd commented on Robey's success by appearing as a cave woman named 'She of the piebald eye', invoking a word that was sufficiently uncommon to seem risqué while also denoting motley or mongrel animal characteristics.[21] This gave Lloyd wide scope to sing about pursuing Robey, whom she called 'He of the carroty crumpet', an apparently innocuous reference to Robey's ginger wig that could easily be inflected with innuendo.[22] The song's inversion of traditional courtship roles also played to Lloyd's hedonistic persona, by suggesting that cave women had been more sexually voracious than their Edwardian descendants. A far more demure and less challenging female response was heard the following April, when Robey performed his sketch in the Easter pantomime at London's Oxford Music Hall, and the soubrette Ada Reeve riposted with a song entitled 'The prehistoric maid' in a revue a few blocks away.[23]

Reports about Robey's sketch and Lloyd's satire appeared in Australia in June, and by the end of the summer a comedian who had just returned from Britain was performing his own version of 'The prehistoric man'.[24] Australians' appetite for cave men remained unsated two years later when George Lauri, a transplanted Londoner who had become one of the country's most popular comedians, toured his version of Robey's sketch through each state.[25] Lauri capped these performances in April 1906 by staging a prehistoric sports meet at the Sydney Cricket Ground to raise funds for theatrical charities (see Figure 12). A large crowd saw participants clad as prehistoric peeps and divided into teams, named the 'Atlantosaurians' and the

12   Prehistoric sports in Australia, *Sydney Mail* (28 March 1906).

'Tyrannosaurians', compete in cricket, rugby, cycling and athletics, under the watchful eye of a club-wielding umpire. The event, films of which were shown throughout the country, reinforced the notion that prehistory was an archaic version of the modern world by having the cave men play violent forms of contemporary middle-class games. Moreover, these cave men appeared alongside stock pantomime characters, yet another suggestion that all of them had been drawn from Britain's past, ever more comprehensively erasing indigenous peoples in favour of a colonial alignment.[26]

The most notable British echo of Robey's sketch was Herbert Beerbohm Tree's 1904 production of Shakespeare's *The Tempest* at Drury Lane. Tree attempted to widen Shakespeare's appeal by incorporating elements of extravaganza and spectacle into his productions. In this one he appeared as the sub-human monster Caliban. The shipwreck that opens the play was enacted on a full-size galleon that pitched as though buffeted by the waves. Such high drama eventually gave way to the cave of the sorcerer Prospero, which was menaced by what *The Times* called 'antediluvian monsters ... Mr Punch's "prehistoric peeps" transferred to the stage', while the good burghers of Derby learned somewhat less floridly that these were 'atrocities modelled on the prehistoric peeps of *Punch* fancy', and similar descriptions appeared as far away as Australia, where Reed's and Robey's versions of prehistory were already popular.[27] The imaginative link between Reed's cartoons and Tree's staging was captured by the *Illustrated London News*, which hired Lawson Wood – a cartoonist discussed in greater detail below – to draw the scene. He depicted Tree costumed as Caliban: a long-haired prehistoric peep with claw-like fingernails, wearing a ragged jerkin and an unkempt Napoleonic beard. He cowers from a group of fanciful dinosaurs that could have been lifted from Reed's drawings.[28] Incorporating such widely known comic monsters was risky, and so it is not terribly surprising to read in the *Era*'s review that Tree's audience 'did not take these strange shapes quite seriously. The stallites seem to regard them as caricatures of a prehistoric period.'[29] Tree's performance may not have been brought to a standstill by laughter, but the report showed that Reed's and Robey's visions of prehistory had conditioned audiences to imagine the era as inherently comic.

The peeps seemed to be omnipresent in the coming years. In 1904, Britain was gripped by the craze for the cakewalk, a frantic dance rooted in American minstrel shows.[30] Reed spoofed this fascination in a cartoon purporting to show the dance's origins: six peeps perform its characteristic high steps in order to avoid the reptiles attempting to devour them.[31] The drawing in turn inspired a dance tune entitled 'The prehistoric cakewalk and two step', whose music was sold in Britain, the empire and the United States under a Reed-inspired cover by the prolific illustrator William George. He drew an axe-wielding peep who winks at the viewer as he cakewalks with a Gibson girl, the epitome of Edwardian beauty, beside a prehistoric sea. The fanciful dinosaurs come from Reed's drawings, while the standing stones that evoke Stonehenge make the scene indisputably British.[32]

A series of political cartoons by Edward Huskinson that appeared in newspapers throughout Britain in 1905 and 1906 under the title 'Prehistoric peeps' eschewed Reed's frivolity. In April 1905, the *Hull Daily Mail* published one that commented caustically on the very public divisions in the Conservative government between traditional free-traders and those led by the Colonial Secretary Joseph Chamberlain, who advocated comprehensive tariffs coupled with preferential rates within the empire. His opponents argued that taxing imported corn, in an era when bread constituted a considerable portion of the working-class diet, would drive up food prices (see Figure 13). They illustrated this at rallies by displaying

13   A prehistoric political cartoon, *Hull Daily Mail* (24 April 1905).

a massive 'big loaf' of bread representing free trade's bounty and a correspondingly tiny 'small loaf' that stood for the meagreness of a tariff regime. Huskinson transposed the dispute, and the political hay that the corpulent Liberal leader Henry Campbell-Bannerman was making of it, into prehistory. Campbell-Bannerman, who would be Prime Minister by year's end, watches with an open-armed gesture of beneficence as a bone-thin peep, whose cap says 'working man', reaches for a 'big loaf' that appears to be held by what the caption calls 'a harmless, useful worm'.[33] Neither perceives that the bread is actually in the tail of a giant, fanged, salivating snake aiming to devour the hungry man. The image suggests that from time immemorial, the people have been served political platitudes and poisoned gifts.

Such pointed cartoons aside, prehistory remained a humorous era that commented on the contemporary world more obliquely. Stage productions that played broadly on prehistoric themes included the comedian George Grossmith's revue *Venus 1906*, which opened that April at the Empire, Leicester Square. The first scene was set in the Park Lane home of a young aristocratic MP, where the statue of Venus in his collection overheard him complimenting a beautiful young woman. The enraged goddess came to life and stole a topical weapon that was 'more powerful than radium', referring to an element that had been much talked about since its discovery eight years earlier.[34] Venus fired this in Trafalgar Square, which was crowded with scruffy Labour Party members, causing a terrible storm in which Nelson's Column and the surrounding buildings collapsed into what the *Stage* described as 'a barren prehistoric waste peopled by those hairy human beings and strange mammals made famous by the drawings of Mr E.T. Reed in the pages of *Punch*'.[35] It was an unsubtle joke that portrayed politically strident members of the working classes as ragged, prehistoric and dangerous, and set them in one of the most widely recognised national and imperial sites. In 1907, Albert Chevalier, a middle-class artiste who had made a career out of playing costermongers, premiered a sketch in which he was transported through past lives to his incarnation as a prehistoric man, an almost identical plot to the play *Peter's Finish*, which was produced in Nottingham that September. British sailors visited a prehistoric island in the 1909 musical comedy *Lollapaloosa* that toured the country, while a benefit matinee two years later at the London Empire entitled *A Pre-Historic Music Hall* saw artistes clad as cave men lampooning political figures. And finally, the 1913 revue *Alice Up to Date* at the

Liverpool Empire concluded with a scene in which Alice and the White Rabbit were transported back to 'Prehistoric Piccadilly'.[36]

Reed was relatively silent amid this booming interest in prehistory and the peeps, though a 1906 cartoon in the *Strand* magazine, depicting him as a cave man carrying a stone axe whose shaft is shaped like a fountain pen, recognised him as the inventor of comic prehistory.[37] Then in March 1907, he helped to create a stage version of his 1894 cartoon 'Prehistoric Lord Mayor's show' for a Drury Lane matinee in aid of a charity whose patron was the Lord Mayor of London (see Figure 14). The Lord Mayor's wife and an entourage of civic-minded women watched performers enact scenes from Charles Dickens, recite poems by Rudyard Kipling, sing songs from W.S. Gilbert and dance the ballet. The afternoon ended with a parade composed of grunting peeps, headed by those clad as the fanciful 'Prehistoric Worshipful Company of Brass Finishers', who marched onto the stage to discordant music and before a backdrop suggesting that London had once been an arid plain. The other livery companies included the Vintners, whose cart lurched on elliptical wheels like a drunkard, while the Mayoress's coach was a wheelbarrow, and a prehistoric City Marshal rode in the Lord Mayor's. In keeping with Reed's knockabout humour, the characters soon began assaulting one another, though, as the *Stage* reported, the 'bodying forth of Mr Reed's "prehistorics" clad in their bearskins, and armed with their tomahawks, clubs and other fearsome weapons, proved rather disappointing', while the *Telegraph* noted more blandly that the battle 'realise[d] in tangible form some of the principal features of these well-known pictures'.[38]

The matinee, in which one of Britain's best-known civic pageants was projected into prehistory, had national impact. This can be gauged by looking, however improbably, at the firm of Stanworth's Umbrellas, which manufactured the most eminently respectable of middle-class accoutrements. Stanworth's was based in Burnley, Lancashire, but sold goods throughout the country by way of an illustrated catalogue. The firm also regularly placed advertisements in both of Burnley's newspapers in which short jokes and mundane anecdotes about umbrellas were illustrated with pencil sketches and cartoons.[39] Then in March 1907, the *Burnley Gazette* printed a report about the Drury Lane matinee, along with a drawing of the prehistoric Lord Mayor's coach.[40] Few if any *Gazette* readers would have travelled to London to see the show, though the editor's decision to report on it indicates an abiding local interest in Reed's cartoons. This was

A prehistoric Lord Mayor's show, *Illustrated London News* (9 March 1907).

equally evident in the twelve advertisements entitled 'Prehistoric pranks' that Stanworth's ran repeatedly on the front page of the *Gazette* and its rival the *Burnley Express* from June until the following spring. Each was illustrated with a cartoon of prehistoric men and women that imitated Reed's style and sense of humour, as did the title (see Figure 15). Successive advertisements depicted cave men inventing tug-of-war, eloping in a dinosaur-drawn sleigh, discovering an umbrella in the Garden of Eden, standing on an umbrella to enact the balcony scene from *Romeo and Juliet*,

15   Prehistoric umbrella, *Burnley Gazette* (25 September 1907).

watching dinosaurs run the Epsom Derby on a damp day and putting umbrellas to many other fanciful uses.[41] One of the final advertisements depicted a young prehistoric couple walking-out under an umbrella while three disappointed suitors and a dinosaur look on, above a text proclaiming that '"survival of the fittest" always proved Darwin's theory to be right'.[42] The advertisement nicely contrasted earlier evolutionary controversies with Reed's vision of prehistory and tied Stanworth's quotidian product to this eternal middle-class Britain.

Stanworth's advertisements, many of which were reused in 1911 in a new series entitled 'Prehistoric Percy', were not unique. Prehistory was invoked in inventive ways to sell everything from tyres to fountain pens. Such advertisements capitalised on Reed's conceit that the prehistoric past was simply an archaic version of the contemporary world, with all its inventions and gadgets. In some advertisements, texts compared new products favourably to their ancient predecessors, while others used cartoons to imagine how a product might have looked in the far-distant past. This gave companies like Avon Tyres the scope to patriotically ally themselves with Britain's history.[43] The company's trademark incorporated a drawing of one of the trilithons from Stonehenge, a symbol of strength and durability, and a semi-mystical evocation of Britain's past. The stones also featured in Avon's advertisements, such as those in 1909 proclaiming that the company's tyres last 'like Stonehenge'.[44] Two years later, this very British firm launched a series of advertisements depicting humorous visions of prehistoric motoring that once again equated the era with other manifestations of British character (see Figure 16).[45] Such advertisements played on the duality of history and evolution. The products being sold had 'evolved' over time through the introduction of new technologies, but the situations in which they were employed and the British people who used them had always existed.

Written invocations of the peeps, like the 1902 review of Rudyard Kipling's *Just So Stories*, which compared these humorous children's tales about animal evolution to Reed's cartoons, drew equivalencies between whimsical views of the very ancient past.[46] The term 'prehistoric peeps' was also used in otherwise serious reports about the discovery of ancient bones from Wales to New Zealand because it brought to mind commonly understood and comforting visions.[47] More than anything, comic prehistory now reflected a sense of British exceptionalism in its suggestion that the nation's ideals and institutions had always been. It was for this reason that books

128 Inventing the cave man

16   Prehistoric motoring, Avon Tyres, *Tatler* (20 March 1912).

about foreign lands and peoples used references to the peeps to denigrate and dismiss such things as the traditional carts on the South Atlantic island of Tristan da Cunha, Moroccan boats, Manchurian peasants and Canada's First Nations.[48] Such usages of the term all suggested Britain's eternal greatness, subtly or not.

These references always reflected the British public's fascination with cave men. Edwardian costume warehouses must have been well stocked with prehistoric furs, wigs and clubs, given the number of amateur dramatic societies that staged Reed-inspired sketches, the balls with prehistoric themes and the historical pageants and parades that included prehistoric scenes (see Figure 17).[49] An analogous fascination is evident in Australia and New Zealand.[50] Reed joined in by helping to decorate the Albert Hall for an artists' ball in 1911, while the following July at a garden fete raising funds for theatrical charities, Robey led a group of performers clad as cave men and cave women under a banner reading 'Prehistoric music hall'.[51]

Cave men also continued to resonate with men in uniform. For instance, the Bedfordshire Imperial Yeomanry recreated a peeps scene for their 1903 Christmas concert, as did a Buckinghamshire territorial unit four years later. The Royal Fusiliers enacted peeps cartoons at an October 1908 fair in South Africa and the following April the crew of the battleship HMS *Prince of Wales* staged a charity concert in Dover that included sailors who 'represented the ancient Britons of *Punch's* Prehistoric Peeps'. They performed before a backdrop on which an imitation of Reed's 1894 cartoon of an archaic Royal Navy had been drawn. The sailors played on this image by casting the peeps as the crew of a Dreadnought, the revolutionary British battleship at the heart of the country's naval race with Germany. The tars also drove a 'Panthard' motor car – a pun on the name of the prominent French motor car maker Panhard et Levassor – that alluded to the use of prehistoric cartoons in contemporary motor car advertisements.[52]

Cave men in public pageants reinforced the popular sense that Reed's version of prehistory was a valid epoch in Britain's past. For instance, one of the figures in Reigate's 1902 cycle carnival was described as Robey's prehistoric man 'pedal[ing] away on an antiquated machine, the wheels of which were made to appear as if cut from stone'.[53] An almost identical report was published five years later about the pageant of British history at a Buckinghamshire cycle parade that featured 'a Prehistoric Man – a regular George Robey conception – on a prehistoric bicycle'.[54] The knowing and jokey idea that Robey, Reed and others were recreating an actual era from Britain's past was acknowledged most prominently and importantly during the Pageant of London, an enormous spectacle staged over four days at the Crystal Palace in June 1911 to mark the coronation of King George V. It opened with 'The dawn of British history', a scene recreating the

130   Inventing the cave man

Mr. Moses Barnett as "Prehistoric Man"

Langfier

17   A prehistoric party-goer: the Jewish feast of Purim in Finsbury, *Tatler* (29 March 1905).

prehistoric past that was, as the *Edinburgh Evening News* stated, 'familiar in present-day pageantry'.[55] It was a pastoral episode set amid stone huts and a replica of Stonehenge – that unparalleled symbol of the nation's eternal existence. At the conclusion of a pagan rite, battered men were seen returning from the hunt, followed by a Celtic army who, according to the event's programme, 'carry stone axes, are clad in skins, have shaggy beards, and look somewhat savage in appearance'.[56] The occasion called for a sober, heroic evocation of Britain's past, but the setting and costumes must have evoked Reed's and Robey's humour in many minds.

Such stage, literary and public evocations also reinforced the notion that prehistory was a masculine era, which in turn supported contemporary gender roles. Female artistes alluded to prehistory, though never as commonly as their male counterparts. Perhaps the most lasting such reference occurred in Marie Lloyd's well-known song of 1910 'When I take my morning promenade', in which she described how her revealing clothing turned young men's heads. The lyrics, which had been written by a man, traced décolletage back to Eve and 'the girls in the prehistoric days [who] each wore a bearskin covering her fair skin'.[57] The song fitted Lloyd's risqué persona, and the lyricist's male fantasies, by implying that women were innately coquettish and hungered for male attention. Another fascinating example that may also have been written by a man was the series of letters from a gossipy, judgemental, envious and status-obsessed middle-class prehistoric woman named Mrs Felicia Stone Axe that appeared in the *Bedfordshire Times* from July 1910 until November 1911. The first letter described a family holiday at Stonecliffe-on-Sea, where the author's husband Reggie played golf with flint clubs, and their children rode mastodons on the beach while unlucky bathers were eaten by sea serpents. Subsequent missives included accounts of the 3rd Stonecroppers, a cavalry regiment quartered near the Stone Axes' home, the difficulty of retaining servants, a whist tournament and a Christmas bazaar raising funds for a new heating system for Stonehenge.[58] The authorial voice reinforced negative early twentieth-century stereotypes about women.

Less salacious or dismissive but equally conservative allusions to prehistory aimed at women included a historical pageant of dolls clad in appropriate clothing from prehistory to the present day staged in 1907 by a Northampton girls' group, and the instructions in how to make a paper doll of Robey's prehistoric man that were published in *Lady's*

Realm magazine three years later.[59] The complacent and demeaning ways in which women were most commonly depicted in prehistory were challenged directly by one of the few cave man cartoons that was definitely created by a woman. It was a now-lost one of 1913 entitled 'The prehistoric argument' that depicted a cave man leaving the family dwelling with his club on his shoulder. When his wife asked why she had to stay behind, he replied that 'woman's proper sphere is the cave'.[60] The image cast men's contemporary resistance to women's demand for social and political rights as the ideas of an outdated and unevolved world.

In the first decade of the twentieth century, rival cartoonists also adopted and adapted, without perhaps improving on Reed's vision of prehistory. In doing so, they gradually supplanted his fame, and sent him into the obscurity from which he has never emerged. Foremost among them was Lawson Wood, whom we have already encountered. His first prehistoric drawings appeared in *Pearson's Annual* at the end of 1902, timing that suggests very strongly that he had been inspired by the popularity of Robey's sketch. Wood was comfortable with the idea of drawing prehistory. He had been born in London in 1878, becoming part of the first generation to have grown up amid a casual acceptance of evolutionary ideas and the conceit that Britain had gone through a comic prehistoric period. He was fortuitously placed to see these ideas translated into images, being just fifteen when Reed's first peeps cartoons appeared in *Punch*. Two years later, after training at the Slade School, Wood began contributing drawings to prominent London magazines.[61]

Catalogues of modern auction houses and print dealers abound with Wood's prehistoric drawings, testifying to the number of cartoons he produced for magazines, postcards, calendars and prints. These images show Wood's heavy debt to Reed, since both men evoked a comic version of prehistoric Britain. But Wood's style was distinguished by rounded, cherubic and vividly coloured characters who inhabit an urban landscape that is markedly different from the one conjured by Reed. Moreover, the subjects Wood covered overlapped and expanded on Reed's satires by including prehistoric visions of such things as motorcycles, divorce, fox hunting, popular songs, courtship and punning proverbs.[62] Wood's drawings were exhibited at a Mayfair art gallery, as Reed's had once been, causing one critic to write that his 'humour has little subtlety … It is extremely clever, however, and each frame contains a broad laugh. Monsters of prehistoric times of fascinating ugliness are comic or ridiculously tragic.'[63] Wood's humour may have been broader than Reed's, but they both used prehistory to comment on the present day.

Wood's cartoons were republished in newspapers and magazines from the United States to New Zealand, just as Reed's had been a decade earlier. These, along with many imitative ones, further cemented the distinctly British conception of comic prehistory throughout the globe.[64] The rate at which Wood produced prehistoric cartoons helps to account for reports like the one in *Melbourne Punch* in 1910 which noted that Reed 'had some fun in him, but he has been beaten at his own game by Lawson Wood'.[65] References to the peeps can be detected over the years, though Reed was similarly displaced in the United States by home-grown cartoonists, who adapted the peeps for the domestic market. American illustrators had begun tentatively exploring Reed's vision of prehistory in the 1890s, though his most notable imitator, Frederick Burr Opper, emerged after the turn of the century.[66] Opper, who as we have seen may well have appropriated an early prehistoric cartoon of Reed's, had trained on two of the most Anglophile American comic magazines.[67] In January 1901, he began publishing a series of cartoons entitled 'Our antediluvian ancestors' in the *New York Evening Journal*, a newspaper that included some of the first comic strips and was owned by one of the nation's largest chains. Opper's style was very clearly indebted to the peeps, while the humour is unmistakably American. The 'Antediluvian' cartoons were intimate and domestic visions of the ancient past because, as Opper explained, 'it seems to me that [prehistoric man's] thoughts must have been very similar to those of the average man of to-day. He, doubtless, considered whether his wife was taking proper care of the dwelling and the children, whether his meals were well cooked, whether he could get the better of some neighbour in a stone-hatchet trade, and whether he could get even with some neighbour against whom he had a grudge.'[68]

Opper adopted this perspective because class did not resonate with Americans, thanks to the country's deep-seated democratic impulses. Opper knew he was on sure ground by sending up such things as marriage and mothers-in-law, a type of domestic humour that was deeply ingrained in British, colonial and American societies.[69] Some of Opper's jokes could not be easily conveyed through visual cues alone, so he included written gags that sounded very much like those that might have been heard on a vaudeville stage. The connection to the dominant form of American comedy was reinforced by the fact that the patter was usually spoken by a pair of cave men drawn by Opper to the side or in the foreground. This gave his cartoons a theatrical quality, and established the men as audience members who observe and actively comment on the scenes.

Opper's urban everymen did not reflect the increasing ethnic diversity of American society and completely erased indigenous peoples from the continent's prehistory, thereby transplanting the idea that Anglo-Saxon society – that of the first colonial settlers – had been eternal. The often unshaven males appear to be manual workers, a more democratic group than Reed drew, while the women are often slightly adipose and seem angered by their husbands' actions. The small and cramped caves in which they live suggest the conditions in America's growing cities. Opper's cartoons were reprinted in newspapers throughout the country, and fifty were published as a book in 1902 in New York and London, where the *Tatler* proclaimed that he was 'assuredly a pupil of Mr Edward Reed of *Punch*'.[70] Opper's success inspired other American illustrators, and cave man comics appeared increasingly frequently in daily newspapers owned by large syndicates.[71] They were often presented as a series or 'strip' of cartoons, a style that offered greater narrative possibilities than the single images that Reed and Wood drew.

Race was central to the American national psyche, making suggestions of consanguinity between apes and humans remained far too contested to be plumbed in cartoons for satire.[72] But the 1904 undergraduate revue at New York's Columbia University shows that this discomfort could be enacted on stage. The show, which was widely reported in the press, included a young man in blackface clad in a gorilla suit, who sang, 'I am the missing link between the monkey and the man... in fact I'm your papa!'[73] Those who preferred associating prehistory primarily with dinosaurs were well served in the era. In 1902, *Tyrannosaurus rex* was first unearthed in Montana by the palaeontologist Barnum Brown. Brown's name, which suggested showmanship and stolidity in equal measure, could not have been more fitting for the man who discovered the species that has become the flag-bearer for all dinosaurs, even though its remains have only ever been found the shadow of the Rocky Mountains. The full skeleton that was put on display at American Museum of Natural History in Manhattan in 1911 is perhaps the most viewed dinosaur of all time.[74]

British and American fascination with the prehistoric world translated most importantly to films, the dominant form of twentieth-century mass culture. The first British films were generally short clips of topical subjects like football matches, horse races and military pageants that were screened in variety theatres. Plot-driven films were more challenging, because of technical limitations and filmmakers' unsure sense of how to move action forward. So they often used chases to literally drive characters from one scene to another and filmed popular stories, knowing that audiences could bridge the huge omissions and compressions that were necessary to distil a novel into as few as five minutes on screen.[75] The long-standing and widespread popularity of Reed's cartoons and their knockabout action helps to explain why in 1905 Cecil Hepworth and Lewin Fitzhamon, two of the most important Edwardian filmmakers, decided to shoot the first film set in prehistory. The four-minute result was entitled, unsurprisingly, *Prehistoric Peeps*. As Hepworth recalled, it was 'a very ambitious film … based upon the work of E.T. Reed of *Punch*, for which all the resources of the works [Hepworth's studio at Walton-on-Thames] were devoted to the building and painting of the wildest of wild animals'.[76] The film was written and directed by the thirty-five-year-old Fitzhamon, who had been a music hall artiste before teaming with Hepworth to make 'sketch' films, whose plots he described as 'very simple, such as could be told in a few words thrown upon the screen'.[77]

Happily, a copy of *Prehistoric Peeps* is preserved in the British Film Institute. It opens as a scientist – wild-haired, grey-bearded and frock-coated – is lowered on a chair into an underground cave. He dances a jig when he sees that its floor is strewn with enormous, ancient bones. He is then evidently overcome by emotion, because he wipes his brow, takes a restorative draught from a flask and lies down to nap. When a dinosaur disturbs his sleep, he shoots at it with a pistol – no laboratory-bound boffin he – before the film cuts to an outdoor scene in which he enters a camp inhabited by women and children made up to look as though they have stepped out of Reed's cartoons. The group flees when dinosaurs with whimsical stripes and spots approach – another direct appropriation from Reed – and the film cuts to a scene of them being pursued down a country road. The film concludes in the scientist's study, where we see him slumped asleep across his desk on which rests a prehistoric femur and a bottle of whisky. The head of a fantastical dinosaur sits atop a cabinet while another has been drawn on a blackboard over the words 'BC 00000' and 'Hepwixosauria', a play on the name of Hepworth's studio. A maid enters and, after failing to rouse the professor, sprays him with soda water. The film fades to black as the shocked scientist awakens.[78]

Film criticism was almost non-existent in 1905, so we have no firm evidence about audience reactions to *Prehistoric Peeps*. But we can speculate that Hepworth knew that the title and subject matter would attract the middle-class audiences he targeted with films that reflected British sensibilities.[79] There is little to recommend *Peeps* today beyond its evident debt to Reed and its role as the progenitor of every subsequent cave man film. It was released in the United States in 1911, though the first British prehistoric film exported to that country – and probably Australia and New Zealand as well – was Charles Urban's 1908 seven-minute feature *The Prehistoric Man*. Its title, which Urban had appropriated from Robey's sketch, showed that he also had a keen sense of how to market his productions. The now-lost film itself owed nothing to Robey, since it told the story of an artist who drew a cave man armed with a stone club. The image came to life, causing havoc on modern streets until the artist drew a dinosaur to chase the brute away.[80]

Just as Opper had reworked the peeps for a domestic audience, American filmmakers soon developed three distinct and often overlapping visions of the prehistoric world in the dozens of cave man films produced before the First World War. The American film industry was coming into its own in the period as nickelodeon machines gave way to purpose-built cinemas that screened longer narrative films with recognisable stars. A series of protectionist regulations limited European imports, while the major studios moved to California and responded to the demands of moral reformers through self-censorship and by promoting their films as uplifting and educational.[81] The vast majority of these earliest American cave man films have been lost, but their plots can be reconstructed from the synopses in the era's many cinema magazines, while still photographs provide concrete clues about how prehistory was depicted. The most startling thing about them is that comedy was not the predominant tone, in contrast to British films or the cartoons in American newspapers. Instead, many early American cave man films aligned prehistory with dominant national narratives, and claimed a pseudo-scientific legitimacy that British showmen had long since abandoned.

Firstly, American filmmakers tied cave men to the metaphysical importance of the frontier in the national psyche. The border between civilisation and the wilderness was a mythic place where resilient, independent pioneers were believed to have forged the country's identity. Tensions between life in the settled east and this idealised western manhood had been explored in early American novels like *The Last of the Mohicans*, in popular stories about 'frontiersmen' like Davy Crockett, and most topically for American audiences, in the well-known life of President Theodore Roosevelt, the sickly Manhattan youth who had gone west and become a man, a war hero and the nation's leader. Large numbers of settlers had reached the Pacific coast by the end of the century, effectively closing the physical frontier, though the idea of such a place resonated in the public imagination. It was evident in long-forgotten literary works like John Corbin's 1907 novel *The Cave Man*, whose protagonist graduates from an elite New England college and enters the automobile business, the industrial frontier of a new century.[82] Roosevelt's biography was more fully matched by Edgar Rice Burroughs's 1912 novel *Tarzan of the Apes* about the orphaned child of English aristocrats – who better epitomised unmanliness in American eyes? – who becomes a masculine paragon when he is raised by apes in Africa.[83] British audiences were unaware of this particularly American myth when *Tarzan* was first published there in Britain 1917 and when the first film version arrived two years later.[84]

Films in this category applied the term 'cave man' to any rugged frontiersman. Plots usually revolved around such men being transported by circumstance, mistaken identity or sudden riches into the urban elite, where they showed that sophistication was a façade and that upper-class women preferred rawer mates. Such films include *The Cave Man* of 1915, based on a hit play in which a society matron wagered that she could introduce a coal stoker – the title character – into her social circle without anyone noticing.[85] The plot device was just as often reversed in films where east coast heiresses found themselves stranded in frontier outposts like logging camps, mining towns and ranches. Intense initial disdain for the male lead's gruff masculinity eventually gave way to a loving realisation that he embodied the ideal. Such escapist fantasies suggested that wealthy and socially advantaged Americans had been enfeebled, while also demonstrating the tensions in a rapidly urbanising society.[86]

Dream sequences were a second common device in American prehistoric films, inspired at least in part by such well-known stories as Washington Irving's 1819 tale *Rip Van Winkle* and Jack London's serialised 1906 story *Before Adam*, about a man who reverts to a simian prehistoric state in his dreams. The plot device was also used in a very successful series of comics by Winsor McCay entitled 'Little Nemo in Slumberland', which followed the dream adventures of a young boy. The cartoons first appeared in the

New York Herald newspaper in October 1905 and were soon featured on board games, dolls, postcards, stage shows, films and an operetta.[87] Dream sequences also attracted American filmmakers because they provided a means of shifting through time, while erasing the continent's rich and varied ancient indigenous cultures.

The most famous dream film is D.W. Griffith's 1912 seventeen-minute *Man's Genesis: A Psychological Comedy Founded on Darwin's Theory of the Genesis of Man*. The unappealing title's portentous claim to scientific veracity masked a plot that very strongly echoed Robey's sketch. The film told the story of Weakhands, a cave man who has been humiliated by Bruteforce, his rival for the maiden Lillywhite. Weakhands wins the contest by lashing a rock to a stick to create a crude weapon with which he defeats his rival.[88] The film's enormous success spawned countless imitators and emboldened Griffith to proclaim that he had created the prehistoric genre, a patently spurious and self-serving assertion that the *Yorkshire Evening Post* debunked in 1914 by publishing its review of *Man's Genesis* under the title 'Prehistoric peeps'.[89]

Griffith's boasting aside, as David Robinson has written, *Man's Genesis* introduced subject matter and techniques that 'gave the cinema new intellectual status'.[90] So the film was widely imitated, not least by Griffith, who frequently remade his own work. In 1915 he directed *Brute Force*, a thirty-three-minute film in which a formally attired man falls asleep while reading a book about primitive hominids. His dream is introduced with an intertitle that proclaims, 'Oh for the good old days of brute force and marriage by capture!' The slumberer conjures up a world in which the members of his fair-haired tribe compete for mates with a dark, heavy-set and hirsute 'womanless' clan. The prehistoric world includes snakes and winged lizards that are introduced with humorous intertitle cards and a *Tyrannosaurus rex*, the most potent symbol of America's ancient past, which is described jokily as 'one of the perils of prehistoric apartment life'. When the dark tribe capture all the women, an intertitle declares that 'defeated by Brute Force, the inventor is compelled to use his brain': that is, to invent the bow and arrow with which he is ultimately victorious.[91] It was a conservative and comforting message about the superiority of slim, fair-haired Western Europeans over the darker heavier tribes that stood in for blacks, ethnic immigrants and indigenous peoples.

Finally, American filmmakers explicitly claimed to have scientifically recreated the prehistoric past. Like Griffith, they invoked Darwin and the concept of the missing link in stories that echoed the evolutionary humbugs spun by P.T. Barnum and other nineteenth-century showmen. Such films included *In the Long Ago* of 1913, which was billed as 'a dramatic tale of reincarnation and prehistoric life, based on recent scientific discovery' that was notable for the rare, perhaps unique, inclusion of North American indigenous peoples.[92] The Afro-centric evolutionary ideas underlying such claims were explicitly explored that August by John C. Hemment, a wildlife cameraman with a strong physical resemblance to Theodore Roosevelt and an equally adventurous résumé. On the eve of an expedition to capture a pair of missing links, he declared to an American cinema magazine that the creature 'is not only prehistoric, but actually exists now, in the wilds of the Congo. He is half-man, half-monkey ... [and] really constitutes the original caveman.'[93] Hemment was actually setting out to film gorillas, but he described them in pseudo-scientific evolutionary language that Signor Farini might well have employed forty years earlier.

Such common but inappropriate appeals to scientific authority annoyed one resident of Wichita, Kansas, who suggested in 1914, 'I have lately seen picture plays purporting to show scenes in the life of cavemen, which certainly give a false idea of the appearance of prehistoric man.' She believed that 'teachers and educators would welcome' a national board charged with ensuring that only appropriate images of the ancient past appeared on screen.[94] Just as in nineteenth-century Britain, this narrow, censorious response to the claims made by self-promoters and publicity agents set the writer at odds with the public's less serious interest in prehistory.

Edwardian audiences did not expect to be taught moral lessons by or learn scientific truths from cave men. They had encountered and absorbed prehistoric imagery and allusions from stages, gramophones, books, magazines, newspapers, films, fetes, costume parties and pageants. These broadcast an increasingly standardised depiction of cave men as ostensibly modern Anglo-Saxons dressed in furs, inhabiting an archaic version of the contemporary world that was beset by fanciful dinosaurs. The vision did not challenge social, gender or ethnic preconceptions. It was so unassailably omnipresent that when the partial skull and jaw of a supposedly British missing link were – again supposedly – unearthed by workmen digging gravel near Piltdown, East Sussex, in early 1912, they were virtually ignored by popular culture.[95] The creature was far too simian. Forty years later it was proved that an unknown forger had intermixed chimpanzee, orang-utan and human evidence. In popular culture, cave men had long since progressed beyond such ape-like missing links to embody innately British qualities. American filmmakers may not have fully adopted this conception of the cave man, but that would soon come.

Notes

1 E.T. Reed, *Prehistoric Peeps* (London: Bradbury and Agnew, 1902); 'Books of today', *Manchester Courier and Lancashire General Advertiser* (2 April 1902), p. 2; 'Prehistoric peeps: the Good Samaritan', *Cardiff Weekly Mail* (31 August 1901), p. 1.
2 Peter Broks, 'Science, the press and empire: Pearson's publications, 1890–1914', in John M. Mackenzie (ed.) *Imperialism and the Natural World* (Manchester: Manchester University Press, 1990), pp. 143–9; G. Philip Rightmire, 'Human evolution in the Middle Pleistocene: the role of *Homo Heidelbergensis*', *Evolutionary Anthropology*, 6:6 (1998), pp. 221–2. See also 'Prehistoric man: discoveries by the Prince of Monaco at Mentone', *Illustrated London News* (10 May 1902), p. 677; 'Are proofs of the descent of man being strengthened?', *Gentleman's Magazine* (January 1903), pp. 43–50; and Nicholas Ruddick, 'Courtship with a club: wife-capture in prehistoric fiction, 1865–1914', *Yearbook of English Studies*, 37:2 (2007), p. 57.
3 'London Pavilion', *Stage* (3 April 1902), p. 17; 'Music hall gossip', *Era* (19 April 1902), p. 20.
4 Both quotations are from 'The Oxford', *Era* (26 April 1902), p. 22; see also 'Players and playthings', *Judy* (9 July 1902), p. 332.
5 Albert Wilson, *The Prime Minister of Mirth: The Biography of Sir George Robey* (London: Odhams, 1956), p. 52.
6 See for instance Erik Allardt, 'Edward Westermarck: a sociologist relating nature and culture', *Acta Sociologica*, 43:4 (2000), pp. 299–306; John F. McLennan, *Primitive Marriage* (Edinburgh: Adam and Charles Black, 1865); and Ruddick, 'Courtship', pp. 45–63.
7 James Harding, *George Robey and the Music Hall* (London: Hodder and Stoughton, 1990), pp. 4–18.
8 'The Tivoli', *Era* (31 May 1902), p. 21.
9 Wilson, *Prime Minister*, p. 51.
10 'Music hall gossip', *Era* (19 April 1902), p. 20.
11 Ibid.
12 Phil May, 'George Robey', *Sketch* (9 April 1902), p. 472; 'Hors d'oeuvres', *Judy* (7 May 1902), p. 222; 'Camberwell Palace', *Stage* (15 May 1902), p. 17; 'London Pavilion', *Stage* (26 June 1902), p. 15; 'Empress, Brixton', *Stage* (10 July 1902), p. 14; 'Stratford Empire', *Stage* (21 August 1902), p. 16.
13 The quotation is from 'The Empire', *Nottingham Evening Post* (26 August 1902), p. 3. See also 'Liverpool Empire', *Stage* (31 July 1902), p. 2; 'Leicester – Palace

Theatre', *Era* (9 August 1902), p. 24; and 'Cardiff Empire', *Era* (6 September 1902), p. 8.
14 'J.W. Sykes', *Yorkshire Evening Post* (17 December 1901), p. 1; 'Gramophone', *Sheffield Daily Telegraph* (10 July 1902), p. 11.
15 'London variety stage', *Stage* (3 July 1902), p. 14.
16 The quotation is from 'The song that made me famous', *Sunday Post* (Lanarkshire) (4 July 1915), p. 5. See also 'Royal General Theatrical Fund', *Era* (22 November 1902), p. 15; 'Palace of Varieties', *Manchester Evening News* (29 September 1903), p. 4; 'George Robey and Phyllis Dare matinees', *Hastings and St Leonards Observer* (27 July 1907), p. 4; and 'Exeter Empire', *Western Times* (23 June 1908), p. 5. Three postcards are in the collection of the National Portrait Gallery: Ax160322, Ax160323, and Ax160324, http://www.npg.org.uk/collections.php (last accessed 6 October 2016).
17 See for instance 'Hull', *Stage* (12 February 1903), p. 13; 'Belfast', *Stage* (2 July 1903), p. 9; 'Ilkeston', *Stage* (24 March 1904), p. 4; 'Stockport', *Stage* (5 January 1905), p. 5; 'Ashton-under-Lyne', *Stage* (19 January 1905), p. 5; and 'Hartlepool', *Stage* (10 January 1907), p. 5.
18 See for instance 'London Hippodrome', *Stage* (3 December 1903), p. 21; 'Gossip', *Stage* (10 December 1903), p. 20; 'Pantomime artists wanted', *Era* (4 November 1905), p. 29; 'Rhyl', *Stage* (27 June 1907), p. 5; and 'Hull Empire', *Stage* (12 March 1908), p. 5.
19 '*Eos and Gwevril*', *Era* (26 April 1902), p. 13; 'A new opera', *Musical Standard* (26 April 1902), p. 261; 'Key notes', *Sketch* (30 April 1902), p. 74. See also Paul Rodmell, *Opera in the British Isles, 1875–1918* (Aldershot: Ashgate, 2013), unpaginated.
20 'A theatrical revue', *Manchester Courier and Lancashire General Advertiser* (28 June 1902), p. 4.
21 'Piebald', in *The Oxford English Dictionary*, vol. VII (Oxford: Clarendon Press, 1978), pp. 834–5.
22 The quotation is from 'The Tivoli revue', *Era* (28 June 1902), p. 16; see also 'Music hall gossip', *Era* (21 June 1902), p. 18.
23 'The medal and the maid', *Playgoer: An Illustrated Magazine of Dramatic Art*, vol. III (London: Greening and Company, 1903), p. 426; 'The Oxford', *Stage* (23 April 1903), p 16; 'The Lyric', *Stage* (30 April 1903), p. 14.
24 'General gossip', *Referee* (Sydney) (11 June 1902), p. 10; 'Personal notes from England', *Register* (Adelaide) (21 July 1902), p. 6; 'Amusements', *Leader* (Melbourne) (2 August 1902), p. 22; 'The theatres', *Sunday Times* (Sydney) (21 September 1902), p. 2.
25 'Her Majesty's Theatre', *Age* (Melbourne) (26 November 1904), p. 16; 'Theatres and entertainments', *Argus* (Melbourne) (12 December 1904), p. 6; 'Mr George Lauri', *Evening Journal* (Adelaide) (20 January 1905), p. 1; 'Special announcement', *Sydney Morning Herald* (6 April 1905), p. 2; 'Amusements', *Daily News* (Perth) (1 August 1905), p. 12; 'Amusements', *Express and Telegraph* (Adelaide) (24 August 1905), p. 2.

26 'Footlight flashes', *Evening Star* (Otago) (11 April 1906), p. 8; 'Prehistoric pranks', *Sydney Sportsman* (11 April 1906), p. 3; 'Prehistoric cricket', *Australasian* (14 April 1906), p. 873; 'The Palace', *Sydney Morning Herald* (23 April 1906), p. 2.

27 The quotations are respectively from 'His Majesty's Theatre', *The Times* (15 September 1904), p. 4; and 'Our London letter', *Derby Daily Telegraph* (15 September 1904), p. 2. See also 'London letter', *Western Daily Press* (15 September 1904), p. 7; and 'The London stage', *Sydney Morning Herald* (26 August 1905), p. 8.

28 '"The Tempest", at His Majesty's Theatre: Mr Beerbohm Tree as Caliban', *Illustrated London News* (24 September 1904), p. 1.

29 '"The Tempest" at His Majesty's', *Era* (17 September 1904), p. 17.

30 Rae Beth Gordon, 'Natural rhythm: la Parisienne dances with Darwin, 1875–1910', *Modernism/Modernity*, 10:4 (2003), pp. 617–56.

31 'The origin of the cake-walk', *Punch* (1 June 1904), p. 385. Reed had satirised the Balfour cabinet a week earlier in 'The ministerial "cake-walk" into the recess', *Punch* (25 May 1904), p. 373.

32 Luke Cavendish Everett, 'The prehistoric cake walk and two step' (London: Hopwood and Crew, 1904), British Library, Music Collections h.3286.cc.(29).

33 'Another prehistoric peep', *Luton Times and Advertiser* (3 February 1905), p. 7; 'Another prehistoric peep', *Leamington Spa Courier* (10 February 1905), p. 2; 'Another prehistoric peep!', *Hull Daily Mail* (24 April 1905), p. 5; 'Another prehistoric peep', *Burnley Express* (13 May 1905), p.7; 'Another prehistoric peep', *Kent & Sussex Courier* (2 March 1906), p. 3. For the tariff debate see also Peter Clarke, *Hope and Glory: Britain 1900–1990* (London: Allen Lane, 1996), pp. 1–39.

34 'The Empire revue', *Era* (21 April 1906), p. 24.

35 The quotation is from 'The Empire', *Stage* (19 April 1906), p. 317. See also 'The Empire', *Stage* (21 June 1906), p. 15; and 'The Empire', *Era* (30 June 1906), p. 19.

36 'The dream of his life', *Era* (13 April 1907), p, 19; 'Peter's finish', *Era* (21 September 1907), p. 25; 'Variety gossip', *Era* (5 June 1909), p. 20; 'Leeds – Barrasford's Hippodrome', *Era* (31 July 1909), p. 7; 'The Hippodrome', *Sheffield Independent* (3 August 1909), p. 8; 'The Hitchins testimonial matinee', *Era* (18 March 1911), p. 28; 'Empire Theatre, the Hitchins testimonial matinee', *Daily Telegraph* (17 March 1911), p. 14; 'Amusements in Liverpool', *Era* (26 November 1913), p. 27; 'Alice Up to Date', *Stage* (27 November 1913), p. 31.

37 '*Punch* makers of today: depicted by each other', *Strand* (October 1906), p. 414.

38 The quotations are from 'Drury Lane matinee', *Stage* (7 March 1907), p. 11, and 'Drury Lane Theatre', *Daily Telegraph* (6 March 1907), p. 9. See also 'Theatrical gossip', *Era* (16 February 1907), p. 14; 'Notes from London town', *Manchester Courier and Lancashire General Advertiser* (8 March 1907), p. 4;

and 'A prehistoric Lord Mayor's show on the stage', *Illustrated London News* (9 March 1907), p. 361.
39 See for instance 'Man-eater', *Burnley Gazette* (9 March 1907), p. 1; and 'What are you looking for my pretty maid?', *Burnley Gazette* (18 May 1907), p. 1.
40 'The Lord Mayor's coach, 10,000 BC', *Burnley Gazette* (9 March 1907), p. 12.
41 See for instance 'Prehistoric pranks', *Burnley Gazette* (12 June 1907), p. 1; and 'Prehistoric pranks', *Burnley Express* (30 November 1907), p. 1.
42 'Prehistoric pranks', *Burnley Gazette* (8 February 1908), p. 1.
43 See for instance 'Avon Tyres', *Illustrated Sporting and Dramatic News* (6 June 1908), p. 565; 'Waterman's ideal fountain pen', *Bystander* (2 June 1909), p. 423; 'Prehistoric Percy's love story', *Burnley Gazette* (8 March 1911), p. 1; and 'Dunlop tyres', *Sketch* (2 April 1913), p. vi.
44 'Avon Tyres', *Bystander* (30 June 1909), p. 641. See also 'Avons for the army', *Bystander* (13 May 1908), p. 358.
45 See for instance 'Avon Tyres', *Illustrated Sporting and Dramatic News* (23 March 1912), p. 165; and 'Avon Tyres', *Tatler* (20 March 1912), p. 143.
46 'Just so stories', *Sketch* (22 October 1902), p. 27.
47 'Prehistoric peeps', *Auckland Star* (26 September 1903), p. 1; 'Prehistoric peeps', *Evening Post* (New Zealand) (11 February 1904), p. 9; 'Prehistoric peeps', *County Observer and Monmouthshire Advertiser* (16 September 1905), p. 6; 'Story book of London', *Oamaru Mail* (4 September 1911), p. 6.
48 *Further Correspondence Relating to the Island of Tristan da Cunha* (London: His Majesty's Stationery Office, 1903), p. 15; Maurice Baring, *With the Russians in Manchuria* (London: Methuen, 1905), p. 33; C.A.F Rhys Davids, 'Il suicidio nel diritto e nella vita sociale', *Mind* (October 1907), p. 617; Charles W. Colby, *Canadian Types of the Old Regime* (New York: Henry Holt, 1908), p. 2; Noel T. Methley, *The Life-Boat and its Story* (London: Sidgwick and Jackson, 1912), p. 17.
49 See for instance 'Bromham open-air bazaar and fete', *Bedfordshire Times* (18 July 1902), p. 8; 'Covent Garden ball', *Era* (27 December 1902), p. 13; 'Masquerade at the Horns', *Era* (31 January 1903), p. 19; 'Flitwick', *Luton Times and Advertiser* (6 January 1905), p. 5; 'The feast of Purim at Finsbury', *Tatler* (29 March 1905), p. 467; 'Wolverton cyclists' parade', *Northampton Mercury* (30 August 1907), p. 6; 'Theatrical gossip', *Era* (2 November 1907), p. 18; 'Annual dinner of the Ampthill Cricket Club', *Bedfordshire Times and Independent* (20 December 1907), p. 12; 'Volunteer prize distribution', *Northampton Mercury* (20 December 1907), p. 5; 'Gloucestershire historical pageant', *Cheltenham Looker-On* (21 December 1907), p. 10; 'Easter at New Brighton', *Era* (17 April 1909), p. 27; 'South Shields carnival', *Era* (18 September 1909), p. 23; 'A children's carnival', *Era* (23 October 1909), p. 28; 'The dance of the four thousand', *Sketch* (1 March 1911), pp. 234–5; and 'Sandgate sports day', *Folkestone, Hythe, Sandgate and Cheriton Herald* (19 August 1911), p. 9.
50 'Fancy dress football match', *Waikato Times* (29 August 1904), unpaginated;

'Students at play', *Star* (Canterbury, New Zealand) (27 June 1905), p. 3; 'Comical cycling coves', *Sydney Sportsman* (12 May 1909), p. 5; 'The pageant', *Table Talk* (Melbourne) (30 December 1909), p. 9.

51 The quotation is from 'London's actor charity', *Sun* (New York) (30 July 1911), p. 5; see also 'London topics', *Hobart Mercury* (31 March 1911), p. 3.

52 'Beds Imperial Yeomanry', *Bedfordshire Times* (11 December 1903), p. 5; 'Volunteer prize distribution', *Northampton Mercury* (20 December 1907), p. 5; 'The busy "beehive"', *Natal Witness* (21 October 1908), p. 7; 'Pulpit and pew', *Natal Witness* (31 October 1908), p. 3; 'Naval variety entertainment', *Dover Express* (2 April 1909), p. 7; 'Prehistoric peeps', *Punch* (11 August 1894), p. 70.

53 'Cycle carnival at Reigate', *Surrey Mirror* (22 August 1902), p. 5.

54 'Wolverton cyclists' parade', *Northampton Mercury* (30 August 1907), p. 6.

55 The quotation is from 'The Pageant of London', *Edinburgh Evening News* (9 June 1911), p. 6.

56 Sophie C. Lomas, *Festival of Empire: Souvenir of the Pageant of London* (London: Bemrose and Sons, 1911), pp. 1–2, the quotation is from p. 2.

57 A.J. Mills and Bennett Scott, 'When I take my morning promenade' (London: Star Music Publishing, 1910), British Library, Music Collections H.3995. zz.(55).

58 'A pre-historic holiday' (8 July 1910), p. 8; 'Prehistoric hop' (19 August 1910), p. 7; 'Prehistoric whist' (21 April 1911), p. 10; and 'A pre-historic bazaar' (3 November 1911), p. 10: all *Bedfordshire Times*.

59 'Northampton working girls' club bazaar', *Northampton Mercury* (29 November 1907), p. 4; 'A new use for lampshades, clever caricatures in crumpled paper', *Lady's Realm* (June 1910), p. 190.

60 'Women's work and enterprise', *Penny Illustrated Paper* (4 January 1913), p. 13.

61 'Mr Lawson Wood', *The Times* (29 October 1957), p. 13.

62 See for instance 'London correspondence', *Sheffield Daily Telegraph* (17 December 1904), p. 8; 'A motto for the Olympia motor exhibition', *Sketch* (13 November 1907), p. 145; 'A decree nisi in the Stone Age', *Sketch Supplement* (20 November 1907), p. 1; 'A prehistoric thoroughbred', *Sketch* (4 December 1907), p. 241; 'A pre-historic Christmas dinner', *Bystander* (11 December 1907), p. 9; and Lawson Wood, *Prehistoric Proverbs* (London: Collier, 1907).

63 'Our London correspondence', *Manchester Courier and Lancashire General Advertiser* (16 November 1905), p. 6.

64 See for instance 'New books and publications', *Argus* (Melbourne) (19 December 1908), p. 6; and 'In the Stone Age', *Topeka State Journal* (Kansas) (16 October 1909), p. 12. Imitative cartoons include 'Prehistoric man had a useful mother-in-law', *Chicago Record-Herald* (12 April 1903), unpaginated; 'News of the day', *Press* (Canterbury, New Zealand) (18 September 1905), p. 6; and 'The prehistoric gee gee', *New Zealand Herald* (22 December 1910), p. 9.

65 'Prattle about people', *Melbourne Punch* (13 October 1910), p. 549.

66 See for instance H.W. Priups, 'Terrible prehistoric combat in New Jersey', *Judge's Library* (September 1896), unpaginated; and Lloyd Mck. Garrison, 'The

work of a great cartoonist', *Cosmopolitan* (September 1900), p. 550. See also 'The science of making gloves', *Washington Times* (16 April 1905), p. 4; 'Horses vanishing from London life', *Semi-Weekly Messenger* (2 April 1907), p. 4; and 'Wanted – a Gillray', *Living Age* (15 October 1910), p. 153.

67 Tom Culbertson, 'The golden age of American political cartoons', *Journal of the Gilded Age and Progressive Era*, 7:3 (2008), pp. 278–81.

68 Frederick Burr Opper, *Our Antediluvian Ancestors* (London: Pearson, 1903), unpaginated.

69 Christine Davies, 'The English mother-in-law joke and its missing relatives', *Israeli Journal of Humour Research*, 1:2 (2012), pp. 12–39.

70 The quotation is from Miles Barry, 'A gossip about books', *Tatler* (22 October 1902), p. 146. See also 'Antediluvian ancestors', *St. Louis Republic* (10 July 1904), comic section, p. 1; and 'Our antediluvian ancestors', *El Paso Herald* (15 May 1915), p. 7C.

71 See for instance 'Our antediluvian ancestors were also some hunters', *Tacoma Times* (20 September 1909), p. 3; and 'Cave man and his cave canem', *Life* (18 December 1913), p. 1126.

72 Rebecca Edwards, 'Politics as social history: political cartoons in the Gilded Age', *OAH Magazine of History*, 13:4 (1999), p. 12.

73 The quotation is from Roscoe C. Gaige, *The Missing Link* (New York: Brooks & Denton, 1904), unpaginated. See also 'Literary notes', *Albany Law Journal* (January 1903), p. 28; 'Students win laurels in opera bouffe', *New York Times* (15 March 1904), p. 9; 'London notes', *New York Tribune* (22 May 1904), p. 9.

74 David Quammen, *The Boilerplate Rhino: Nature in the Eye of the Beholder* (New York: Oxford University Press, 2000), pp. 121–5; 'If that dinosaur were strolling in State Street', *Chicago Tribune* (26 July 1907), p. 5.

75 Roger Sabin, 'Ally Sloper on stage', *European Comic Art*, 2:2 (2009), pp. 205–25; Bryony Dixon, 'The ancient world on silent film: the view from the archive', p. 28; and Pantelis Michelakis and Maria Wyke, 'Introduction: silent cinema, antiquity and "the exhaustless urn of time"', pp. 1–5: both in Pantelis Michelakis and Maria Wyke (eds.), *The Ancient World in Silent Cinema* (Cambridge: Cambridge University Press, 2013).

76 Cecil Hepworth, *Came the Dawn: Memories of a Film Pioneer* (London: Phoenix House, 1951), p. 69.

77 Denis Gifford, 'Fitz: the old man of the screen', in Charles Barr (ed.), *All Our Yesterdays* (London: British Film Institute, 1986), p. 317.

78 Cecil Hepworth (dir.), *Prehistoric Peeps* (1905), British Film Institute. See also 'Hepwix', *Era* (20 January 1906), p. 33.

79 'Prehistoric peeps', *Optical Lantern and Cinematograph Journal* (April 1905), p. 142; Andrew Higson, 'Introduction', in Andrew Higson (ed.), *Young and Innocent? The Cinema in Britain 1896–1930* (Exeter: University of Exeter Press, 2002), pp. 2–8.

80 'Hanley', *Stage* (22 October 1908), p. 4; 'The prehistoric man', *Moving Picture*

World (14 November 1908), p. 384. See also 'Entertainments', *Western Mail* (Australia) (12 December 1908), p. 38; and 'Tivoli theatre', *New Zealand Herald* (17 February 1909), p. 12.

81 David Robinson, *From Peep Show to Palace: The Birth of American Film* (New York: Columbia University Press, 1996), pp. 101–58.

82 John Corbin, *The Cave Man* (New York: Appleton, 1907).

83 David Leverenz, 'The last real man: from Natty Bumpo to Batman', *American Literary History*, 3:4 (1991), pp. 753–65.

84 See for instance 'Recent fiction', *Sphere* (15 September 1917), p. 214; 'Alhambra', *Bystander* (24 December 1919), p. 1036; and 'Monkeying with a peer!', *Sketch* (14 January 1920), p. 81.

85 'The cave man', *Moving Picture World* (20 November 1915), p. 1562.

86 See for instance 'A cave man wooing', *Implet* (11 May 1912), p. 2; 'Trial marriage', *Variety* (1 November 1912), p. 25; 'The primeval test', *Motion Picture News* (1 November 1913), p. 44; and 'The man with the hod', *Motion Picture News* (20 May 1916), p. 2999.

87 Tom W. Hoffer, 'From comic strips to animation: some perspectives on Winsor McCay', *Journal of the University Film Association*, 28:2 (1976), p. 24.

88 Stella MacHefert, 'Man's genesis', *Motion Picture Story Magazine* (August 1912), pp. 100–6.

89 'Prehistoric peeps', *Yorkshire Evening Post* (19 February 1914), p. 4. See also Chas. L. Gaskill, 'Before a book was written', *Motion Picture Story Magazine* (April 1912), pp. 25–35; 'The miser's reversion', *Motion Picture News* (21 March 1914), p. 52; 'In wildman's land', *Motion Picture News* (19 December 1914), p. 96: 'Biograph re-issues six more Griffith pictures', *Motion Picture* News (26 June 1915), p. 63; and Judith C. Berman, 'Bad hair days in the Paleolithic: modern (re)constructions of the cave man', *American Anthropologist*, 101:2 (1999), p. 289.

90 Robinson, *Peep Show*, p. 130.

91 D.W. Griffith (dir.), *Brute Force* (1915). See also 'Two famous biographs', *Moving Picture World* (9 October 1915), p. 285.

92 'In the long ago', *Moving Picture World* (10 May 1913), p. 575. See also 'The revelation', *Moving Picture World* (4 October 1913), p. 67; 'Race memories', *Moving Picture World* (25 October 1913), p. 350; Charles Ingram, 'Film gossip', *Illustrated Films Monthly* (February 1914), p. 338; and 'The miser's reversion', *Motion Picture News* (21 March 1914), p. 52.

93 J.C. Hemment, 'My search for the missing link', *Motion Picture Magazine* (August 1914), p. 77.

94 'What improvement in motion pictures is needed most?', *Motion Picture Story Magazine* (February 1914), pp. 93–4.

95 John Evangelist Walsh, *Unravelling Piltdown* (New York: Random House, 1996), passim.

6

Cave dwellers of Flanders: the First World War

By 1914, prehistory was a frequently evoked and popularly accepted, if chronologically indistinct, epoch in Britain's past that fell somewhere between the age of the dinosaurs and the Roman invasion in 55 BCE, which marked the advent of written records. It was a mythic time in which Britons fought fanciful reptiles and consorted with ancient heroes like Queen Boudicca. It was also a comic era in which the contemporary world was presented in archaic and shambolic terms. No one claimed that this was the past as it actually happened, but it was comforting to think that it was, because cartoons, sketches and films broadcast the subtle but clear message that British ideals had existed from time immemorial, giving them a legitimacy that those of other nations lacked. The conceit resonated throughout the empire, where colonists clung to British identities and ignored the ancient cultures amid which they had settled. Americans had transformed cave men to fit their more multi-ethnic settler society, with its deep antipathy to social class distinctions. Though the United States had been born in a rebellion against Britain, prehistoric scenes comfortingly proclaimed the supremacy of a shared Anglo-Saxon heritage.

In Britain, allusions to this comic prehistoric era abounded in 1914. In April, Avon Tyres used cave men cartoons to advertise its products, while the dance 'Prehistoric zig zags' appeared on gramophone records and as sheet music that was sold under a vivid, full-colour drawing of a cave man and woman by William George, echoing the image he had created for 'The prehistoric cakewalk' a decade earlier.[1] In May, a Unionist MP described a statement by Herbert Samuel, the president of the Local Government Board, as being 'like a prehistoric peep' for its retrograde understanding of modern conditions.[2] May also saw the revue *Go-Ahead* at the Nottingham Hippodrome, which recast 'familiar music hall turns in a burlesque prehistoric setting', as had been done many times in the previous decade.[3] Dancers clad as peeps were seen at a Royal Botanic Garden fete in June, while archers aimed their arrows at a huge dinosaur at a Buckinghamshire carnival in July.[4] Historical pageants throughout the country also often commenced with scenes set in prehistory, reinforcing a teleological sense that the era was the foundation of Britain's current global dominion. Such events included the 'stirring yarns from history' acted in January by Cardiff Wolf Cubs, which began with scenes of a 'prehistoric boy' and progressed through time to scenes representing members of the Scouting movement.[5] The summer fete at Bourneville, the model village near Birmingham for workers at Rowntree's factory, presented scenes of British history including Robin Hood and a sketch entitled 'prehistoric peeps' depicting how chocolate had been sold in the ancient past. It was interrupted by a fanciful dinosaur.[6] By pairing the scenes, the workers made them equivalent evocations of the British national character. It did not matter that the first story had been handed down for centuries and the other was little more than two decades old.

British audiences also watched American cave man films in the first half of 1914, most notably *Through the Ages*, in which a boxer was hit on the head and dreamt of fighting in the Stone Age. Advertisements in Britain called it, almost inevitably, a 'screaming prehistoric peeps comedy', though it was actually a topical comment on the sport of boxing's well-publicised search for the 'great white hope'.[7] This was a racist quest for someone to beat the Texan Jack Johnson, who in December 1908 had become the first black man to win the heavyweight world championship. American cartoons depicted Johnson as an unevolved gorilla as a way of articulating racist anxieties about his seeming invincibility and the more existential threat that this represented to white society.[8] At least one British showman made the association between the film and the controversy overt, by pairing *Through the Ages* with a film of the bout between Britain's 'great white hope', Bombardier Billy Wells, and Australia's contender, Colin Bell.[9]

Pre-war visions of Britain's prehistoric past were rarely so charged. They culminated with the film that Seymour Hicks shot at the end of July 1914 on the Dorset coast. Hicks, one of the most popular Edwardian musical comedians, played King Mugslot, the aged ruler of an ancient land. His costume paired a Classically inspired tunic and sandals with a spiky wig and beard that were clearly derived from George Robey, while his subjects were clad as peeps in imitation of E.T. Reed (see Figure 18).

Prehistory in film: Seymour Hicks as King Mugslot and Jessie Fraser as Coral, *Sketch* (29 July 1914).

The plot, concerning Mugslot's attempt to marry a girl named Coral, borrowed heavily from established pantomime and music hall gags, like the couple's visit to an archaic Selfridges – a familiar send-up of London's most modern department store – and a car number plate that read 'BC IOU', a gag derived from the 'T8' one that the comedian Harry Tate used on his private motor car. The reluctant bride fell in love with a man who washed ashore on a raft. He was young and handsome and belonged to a more advanced species. An enraged Mugslot reprieved the suitor from death on condition that he share his sailing technology. The film ended with Mugslot's comeuppance as he lifted his bride's veil at the altar only to find that she was not Coral, who had escaped with her love. Still photographs are all that remain of a film that was made a month before the war began and released only the following spring.[10]

American cave man films were exported to Australia and New Zealand, where local artistes continued imitating Robey's prehistoric man sketch, and manufacturers and purveyors of a variety of goods incorporated images and allusions to the era in their advertisements.[11] Much the same happened in the United States, though cave men continued to compete with dinosaurs in the popular imagination. The American fascination was reflected on screen by Winsor McCay, a popular cartoonist and vaudeville performer who took the stage clad as an animal trainer to give a performance in which he drew quick images of his most famous creation, Gertie the Dinosaur. In early 1914 McCay transferred the character to a partially animated film, which opened with scenes shot in the dinosaur hall of Manhattan's American Museum of Natural History, the institution most associated with these ancient reptiles. It was an unsubtle way of appropriating scientific credibility for his friendly brontosaurus.[12] Edgar Rice Burroughs once again played broadly on the American man myth with his story *The Eternal Lover*, which was serialised in magazines. In it, a prehistoric hunter was transported 100,000 years through time to fall in love with a Midwestern girl who is visiting Burroughs's most famous creation, Tarzan, in Africa. She cannot resist the time-traveller's primal manliness, and the pair elope to the Stone Age.[13]

Such evidence indicates that over the previous decade, prehistoric cartoons, sketches, films, fiction and advertising had created a common understanding of the ancient past throughout the English-speaking world. Shared references in turn gave men in uniform, newspaper correspondents, propagandists and civilians an argot with which to articulate their experiences. This was evident at the start of the First World War, when amateur poets extolling the British cause in newspapers sometimes used prehistoric imagery. These would-be Kiplings included a person whose work appeared in the *Manchester Evening News* in late September 1914, denouncing the

Kaiser's eagerness for war as 'the fierce and ruthless greed / the lust for world dominion; the caveman's savage creed'.[14] The following January, a poem in the same newspaper with the pessimistic title 'Armageddon' attributed the conflict to 'some squat and hairy cave-man in a state of slow transition'.[15] Journalists mined prehistoric imagery equally keenly in bellicose declarations like 'we are to revert to the regime of the cave-man. We are to fight tooth and claw like the apes who aeons ago survived by reason of, and in proportion to, their strength and ferocity', which suggested that British men who joined up were reconnecting with their atavistic, though implicitly human, prehistoric selves.[16] Other prehistoric phrases challenged high-minded statements about the war's aims: 'Civilisation say you? The civilisation of the cave-man rather, without the cave-man's excuse.'[17]

Baser and more simian instincts were attributed to the Germans. After a 1915 air raid on Norfolk, which breached the Hague Conventions forbidding the bombardment of undefended towns, the *Birmingham Gazette* denounced 'Zepplinism' as 'defy[ing] the moral judgement of the human race. It sacrifices the good name of the German people in order to prove that German Kultur joins its white hands with the cave-man in the Stone Age.'[18] Cartoons broadcast the message more forcefully, beginning with the image that the Australian-born illustrator Will Dyson published in the *Daily Herald* newspaper on 1 September 1914. Dyson depicted a club-carrying simian standing fawningly beside the Kaiser, who wears his military and imperial regalia. They are watching Louvain burn – German soldiers had razed the ancient Belgian city and killed many of its civilian inhabitants the previous week – an action of which the gorilla is proud. The cartoon's title quotes the Kaiser, 'remember your illustrious forbears', suggesting that this war crime represented innate German values.[19]

Dyson again captured this sense that a society that had produced great art, literature and music had regressed to an atavistic sub-human state in a 1915 cartoon in which a hairy, brutish and simian Neanderthal is transfixed by a professor who holds up a vial of some noxious substance and declares, 'together, my Herr Cavedweller, we should be irresistible!'[20] The image commented on the German army's large-scale use of poison gas that January. Though both sides would eventually deploy chemical weapons, Dyson expressed the revulsion with which civilians greeted this symbol of attrition warfare. To Dyson, Germany had formed an alliance between a devotion to modern science and menacing, simian prehistoric characteristics. Like *Punch*'s 1861 'Mr Gorilla' cartoon, which Dyson may well have known, the unevolved animal in his image is addressed with a gentleman's title, suggesting that it has invaded a world that had heretofore been restricted to civilised, modern creatures and chivalric notions of battle. The animal is clearly submitting to the unscrupulous German scientist, who sees that unleashing the fierce brute is a way to satisfy his own lust for victory and power.

Dyson's cartoons were also published in the United States, a still-neutral country where depictions of Germans were initially less stridently negative.[21] For instance, a February 1915 cartoon from *Life* magazine showed a cave man with a hatchet in his belt, holding his club like a rifle. Next to him is an empty German uniform, drawn to resemble a suit of armour, on which are written the words 'honor, patriotism, heroism, bravery'. The cave man is not ferocious, menacing or simian. He is a Cro-Magnon derived from comic depictions of prehistory. This, combined with how the empty uniform is posed, suggests that German soldiers have set their ideals aside without yet forsaking their humanity.[22] Prehistory also provided American cartoonists with ways to explore the deep divisions about the nation's attitude to the war. These tensions are evident in a cartoon that appeared in a Chicago newspaper a few days before the presidential election in November 1916. It depicts the Republican candidate Charles Hughes as a hirsute and heavy axe-carrying cave man who has pummelled a young woman representing American 'Peace and Prosperity'.[23] It was a harsh comment on Hughes's interventionist stance, suggesting that it was not some comic version of Stone Age courtship, but an immense threat to the nation. The image also deliberately contrasted Hughes with his opponent, the incumbent Woodrow Wilson, who was running under the slogan 'He kept us out of war'.[24] Wilson was returned to the White House, but nonetheless sent American troops to fight the Entente powers in January 1917.

Thereafter, Germans were regularly compared to prehistoric beasts in the United States. Most famously, a widely reproduced recruiting poster featured a snarling open-mouthed gorilla wearing a pickelhaube, the characteristic German spiked helmet, emblazoned with the word 'Militarism'. In one hand the animal holds a huge club on which is written 'Kultur', a term that had heretofore related to the country's artistic achievements, and in the other it clutches a swooning, bare-breasted maiden. The shattered remains of Europe lie in the dark background as the monster steps towards the viewer and onto American soil. The slogan advises reader to 'Destroy this Mad Brute', before, as the image unequivocally suggests, he sexually, culturally and physically assaults civilised ideals.[25] The drawing, which screamed that German victory would impose simian brutality on Western society, was rooted in mid-nineteenth-century evolutionary cartoons and the inter-species sexual and racial threats they represented. Other government-produced images proclaimed that German soldiers had wholly rejected idealism for savagery. These included the advertisements for war bonds that appeared in newspapers and magazines throughout the country in early 1918. They bore the provocative title 'The destruction of civilization', and included a photograph of a 'trench club', a gruesomely effective implement made by driving spikes into a heavy wooden mallet, described as something that 'might be the weapon of a savage cave man of five thousand years ago. It is in fact the weapon with which German soldiers "finish off" enemy wounded who have fallen on the battlefield.'[26] The clear but spurious inference was that American troops adhered to idealistic notions of warfare.

As the war progressed, corollary allusions continued to portray the cave man qualities of Britain and her allies in wholly positive ways. Reed, Robey and others had depicted British cave men standing up to more simian rivals and acting cheerfully as their institutions and ideals collapsed into chaotic, comic violence. Such cartoons, songs and sketches had suggested that Britain and British ideals, like duty, courage and sacrifice, were innate and timeless. The idea that these eternal traits had been reawakened by the war was articulated by the Scottish mythologist Lewis Spence in a short article that was published in a Scouting annual near the end of the war. Spence expected the boys at whom his work was aimed to be conversant with images that dated to their own parents' youths, beginning:

> everyone is more or less familiar with those amusing pictures that appear from time to time in the pages of our comic journals, which are supposed to caricature the life and customs of prehistoric times. In the drawings of genial humorists we behold skin-clad individuals making agonised efforts to get out of the path of scaly monsters, or pressing home the point of a bad joke with the aid of a stone axe. These pictures, one feels, have done much more to form the popular idea of what 'prehistoric' times were like than all the efforts of learned professors.[27]

The rest of the article substituted this comic imagery with a wartime metaphor, noting that in the Stone Age 'the constant struggle for existence, the everlasting search for food, did not give much scope for sport as the modern boy knows it. But of sport of another kind there was plenty and to spare – a kind in which failure often meant a sudden and dreadful death.'[28]

British authors had often described war over the previous decades through metaphors drawn from sport and the public school cult of games, but Spence suggested that the analogy had a far more profound historical basis.

Despite such efforts to assert serious cave man qualities, the idea that prehistory was inherently comic was not easily dislodged from the imaginations of men in uniform. They made references to the era to express their experiences of the war, much as an earlier generation had done about the fighting in South Africa. It could not have been easy to see the war in these terms, given the horrors men faced. This struggle was evident in an open letter from the London actor Thomas Treherne that appeared in the *Era* in 1916. Treherne, who was serving as a junior officer in Flanders, reported sensing that 'I grew more of an animal – more of a beast, I might say, when I think of the loathsome horrors which I came to regard with indifference – I found my artistic faculty dying away … I became less of an actor and more of a primeval savage day by day.'[29] Treherne clearly found neither humour nor heroism in the links between war and prehistory, sensing instead that he was reverting to a more 'savage' phase of human evolution. The opposite was true with Alex McGhie, a footballer who had played for Blackburn and Liverpool. He described a match near the front line – sport was an important means of maintaining troop morale – as 'prehistoric football'. In McGhie's eye, the war-scarred physical surroundings, impromptu pitch and players clad in army boots and khaki trousers brought to mind the archaic games that had so often been drawn in pre-war cartoons and enacted on stage and in pageants.[30]

McGhie's letter, which was published in a Lancashire newspaper, also showed that prehistoric imagery could soften the war's horrors for civilian audiences. If young men in uniform still spent their spare time playing football, a game whose public school origins had given way to broad national appeal, then British idealism survived. A similar sentiment is evident in the subtitle to a late 1915 drawing in the *Graphic* of artillerymen moving guns in the half-light of dawn that compares the weapons to 'a great living prehistoric monster feeling for its prey', suggesting somehow that the war was a lark.[31] Such imagery became especially popular in mid-1916 as trench lines stabilised, the toll of deaths increased dramatically, and commanders sought a decisive strategic breakthrough. Hopes were high that tanks, a weapon that the British had developed in secret, would win the war at a stroke. These cumbersome and ungainly machines were immediately compared to comic prehistory because they would not have looked out of place in one of Reed's drawings. Furthermore, the massive advance in battlefield technology that they were believed to represent recalled the myriad cartoons, sketches and films in which Cro-Magnon cave men had used their brains to defeat larger, aggressive, unevolved rivals.

The association appears to have first been made by William Beach Thomas, a correspondent for the *Daily Mail* and other publications who was noted for enthusiastic war-boosting and florid prose. In a widely reprinted report of mid-September 1916, he likened tanks moving on a moonlit night to 'blind creatures ... from the primeval slime' and most especially to Lewis Carroll's Jabberwock.[32] Within a week, British frontline troops had absorbed and adapted Beach Thomas's description. The *Chester Chronicle* quoted a British officer who described tanks as being 'like prehistoric monsters, you know, the old ichthyosaurus' before, as the reporter noted, laughing 'in a queer way at some enormous comicality' about how the machines chuffed and clanked with seeming impunity through the soldiers' troglodyte world of trenches, mud and craters.[33] To this observer, tanks were far closer to Reed's comic dinosaurs than to Carroll's poetic monster. Canadians likened tanks to mammoths, the giant prehistoric mammal that epitomised American grandeur, and the Germans who tried to fight them to inadequately armed prehistoric hunters.[34]

References to Reed's drawings became even more common after films and photographs of tanks in action were released to civilians in late 1916. Advertisements called the new weapon 'an amazing creature worthy of "prehistoric peeps"' and 'a monster out of prehistoric peeps' among other things.[35] New Zealand's official war correspondent invoked prehistory to convey the experience of life at the front at least once, and the country's soldiers compared tanks to the hansom cab that Reed had depicted twenty years earlier.[36] At war's end, Canadians described tanks at Bovington Camp, Dorset, as 'disporting themselves like huge prehistoric monsters clothed with thunder'.[37] Such associations combined optimistic resolve with a schoolboy delight that comfortably cloaked the grim reality of the trenches.

These accounts showed that men in uniform found a language with which to spoof the war in Reed's comically violent but archaically civilised world. The vast majority of men had never set out to be soldiers. The tenacity with which many clung to their pre-war identities included taking a bemused attitude to unthinking military discipline and hierarchy. The wartime world in which they found themselves could have been conjured by Reed, with its mixture of senseless violence, scarred and deforested landscapes, and a military that reproduced and codified Edwardian class structures. So it was almost inevitable that allusions to prehistory appeared repeatedly in the satirical 'trench newspapers' published by units from throughout the empire. These ranged in quality from a few handwritten pages to those that were professionally typeset and bound. They mimicked *Punch* and similar comic magazines by sending up superior officers, strategy and daily life, and satirised official war correspondents to show how absurdly distorted their reports were.

In late 1916, the *BEF Times*, the most famous trench newspaper (originally published as the *Wipers Times*), spoofed Beach Thomas in a report about tanks that described how 'in the grey and purple light of a September morn they went over. Like great prehistoric monsters they leapt and skipped with joy'.[38] Tanks inspired such humorous reactions because they were British, evoked the nation's comically imagined eternal characteristics, and carried with them a comforting sense that they might bring the war to a speedy close. The jolly, self-mocking tone extended to the way in which soldiers referred to themselves as the 'cave dwellers of Flanders', lamented their low pay as 'prehistoric dough', discussed 'cave man' tactics for courting women while on leave and called war 'the prehistoric manner of settling differences'.[39] By contrast, soldiers' tones changed abruptly when describing the German munitions that terrorised, maimed and killed them. These were a 'vast horde of unearthly creatures, chief of which and most ferocious is "the minenwerfer" [a widely used mortar] – prehistoric but lately granted a new lease of life'.[40] Like Will Dyson's cartoons, such phrases showed that soldiers believed their foes had reverted to pre-human savagery.

By 1914, prehistory had become a predominantly visual era. So it is not terribly surprising to find that Reed's cartoons were imitated by artists in uniform. The images appealed especially, as the reaction to tanks has forewarned us, to men who grappled with new military technologies. Such weapons were often hastily designed with little consideration for aesthetics or comfort, assembled from roughly finished parts and deployed on the battlefield without having been thoroughly checked for safety. As a result, wartime inventions often looked prehistorically unrefined with the added danger that if they did not work properly – a fairly common occurrence – their operators would be maimed or killed. The first such cartoon in a British trench newspaper was 'The war in the air', which appeared in September 1916 in the *Gasper*, a publication produced by the Royal Fusiliers. It depicts a prehistoric version of the Royal Flying Corps, in which wicker baskets akin to the ones used by military balloonists in earlier wars have been strapped to enormous birds. The image places military aeroplanes, the most radically advanced wartime technology, in historic and prehistoric contexts.[41]

A more imaginative, unknown artist produced 'Hospital evolution', which appeared in the following month's edition of the *Searchlight*, the newspaper of a British military hospital (see Figure 19).[42] The artist's cave men are essentially advanced stick figures, though he clearly attempted to portray the stretcher bearers, doctor and wounded soldiers as prehistoric peeps, with gangly limbs, shock hair and fur garments. They inhabit a rocky and sere wilderness, in which a cave serves as the hospital. It is a Reed-like prehistoric version of the unit's Manchester base. Fanciful dinosaurs and huge bat-like creatures look on, suggesting an unfriendly and inherently dangerous world. The subtitle dates the scene to about 3500 BCE, showing the chronologically carefree way in which prehistory was inserted into Britain's past. The static image lacks much humour or dramatic tension, and yet it provides telling insights into the soldiers' experience. It is a humorous critique by an artist who had been thrust into a violent, wild and physically scarred world where injury and death were ever present and medical care seemed as ineffective and obsolete as that which had been available to cave men.

Prehistoric military medicine, *Searchlight* (October 1916). 19

A very similar spirit infuses a pair of cartoons that appeared in mid-1917 in the *Gnome*, a newspaper published by a British school of military aeronautics that was stationed in Egypt. The unknown artist convincingly recreated Reed's style and sense of humour. In the first image, he drew an outdoor classroom, enclosed by a palisade of standing stones in which young men clad as peeps are learning the elements of flight. The arrangement of the pupils' benches and the barren landscape they inhabit suggest it is an adaptation of Reed's 1894 cartoon 'A night lecture on evolution' to fit wartime experience (see Figure 7). The aeroplane crashing to the ground in the distance is a fatalistic, comic commentary on the very hazardous duties that young men undertook in machines whose rudimentary Heath Robinson appearance and mechanical unreliability suggested that they belonged to an archaic version of the modern world. The humour is repeated in a second cartoon, depicting another prehistoric aircraft (see Figure 20). Here the fuselage is an assembly of sticks that have been lashed together, and the wooden wheels recall those of pre-war prehistoric bicycles and motor cars, while the skids have been fashioned from the horns of giant

20   Prehistoric battlefield tactics, *Gnome* (August 1917).

mammals. A bombardier drops heavy stones on the enemy below, while anti-aircraft gunners reply in like manner by launching rocks from huge catapults. The presence of a giant dinosaur, which eyes the terrified enemy infantry hungrily, is explained in the subtitle, which reads 'an aeroplane co-operating with a tank'. The dinosaur therefore reflects written comparisons between tanks and ancient monsters that had first appeared a year earlier. The cartoon's simple visual gag conceals two more subtle messages. The gimcrack aeroplane is yet another grim comment on the notoriously short life expectancies of First World War flyers, while also satirising the necessity of cooperation between aerial and armoured warfare, entirely novel and sophisticated tactics that were crucial to deploying new military technologies.[43]

The battlefield was a very masculine environment. But women were also crucial to the war effort. They took over previously male jobs in factories and offices and worked in military hospitals and convalescent homes. One of the few early prehistoric images of cave women to be produced by a woman, it seems probable, accompanied an article by a volunteer at a Scottish military hospital (see Figure 21). It illustrated a piece written by a titled woman who had volunteered as a nurse. She loathed the imperious matron under whom she worked, wondering, 'as the fluff at the back of my

An early depiction of cave women, *Thistle: Scottish Women's Hospitals for Foreign Service Souvenir Book* (Glasgow: John Horn, 1916).

neck stiffens with ancestral stirring of a by-gone scruff, what was the prehistoric origin of my enemy?'[44] The cartoon conveys the article's *infra dignitatem* sentiment, by depicting the matron as an imposing, club-carrying taskmaster revelling in the power conveyed by her position. The author is younger and slimmer and has topped her smart fur garment with a feather in her hair, such as she might have worn to a society ball. She sneakily thumbs her nose at her enemy. The cartoon's humour was based in social class – the author bristled at submitting to the authority of someone she would normally have considered her social inferior – and therefore played in the same register as Reed's images.

While trench newspapers were aimed primarily at men in uniform, they also circulated on the home front. Their familiar satirical tone would have suggested that humour prevailed in the trenches and appears at times to have shaped public perceptions of the war. For instance, a 1915 trench newspaper produced at Gallipoli included a drawing of a digger, as Australian soldiers were known, being visited by a cave man. The troops at Gallipoli inhabited a sere and barren environment that would easily have evoked prehistory. In a long poem, the soldier explains that the cave man should feel at home, because:

> we're living still in caves, sir, dug mostly out of clay
> We call them trenches, dug-outs, saps; but call them what we may
> They are made to hide our skins, just as in your heathen day.[45]

The Gallipoli campaign, which lasted from April 1915 to the following January, was almost immediately accorded mythic status as the forge of modern Australian identity. It was commemorated with a parade through the streets of Adelaide in October 1916. Alongside the rather predictably patriotic participation of John Bull, pipers, marching bands, soldiers, civic dignitaries, union leaders and parliamentarians walked a pair of huge fanciful prehistoric animals, dubbed the 'Kangarusaurus' and the 'Camelephant', that could have been lifted from Reed's cartoons. They were also topical comments on the war, in that the first was an ancient inhabitant of the diggers' homeland and the latter an evocation of the place where many had served and died. The beasts were accompanied by ten cave men, of whom one rode a Stone Age tricycle with a ram's head for a steering wheel, and another periodically clubbed those close to him.[46] The parade proved that the cave man had become a national and imperial symbol of wartime indomitability.

Prehistoric images from trench newspapers and soldiers' letters resonated on the home front partly because similar images were seen frequently by civilians. British advertisements continued using cave men cartoons, like those in 1915 for Stanworth's Umbrellas, the Lancashire firm that had first employed such images eight years earlier. These updated images concerned 'Prehistoric Percy', a cave man who headed a troop of prehistoric volunteers while wielding an umbrella that was now patriotically emblazoned with the Union Jack. Subsequent advertisements showed Percy in military settings, for instance using an umbrella as camouflage, and as a bomb-proof shelter.[47] Less overtly militaristic images appeared in advertisements for the kitchen cleanser Panshine, whose Reed-derived cartoon promised to brighten pots and pans with the gentlest rubbing, unlike Stone Age antecedents.[48] Glasgow art students created a prehistoric village for a 1916 Christmas bazaar, and as coal supplies ran low in mid-1918, the government produced cartoons that reminded British householders that cave men's fires had been far more efficient than the average modern hearth.[49] Similar cartoons were used to sell boots in New Zealand, where at least one prehistoric hockey match was staged, while in Australia cave men promoted patent medicine and suitcases, all on the basis that these products improved on their ancient predecessors.[50]

Prehistoric imagery was equally prominent in the United States during the war. Newspapers and magazines featuring drawings of cave men proclaimed that 'Nujol', a petroleum-derived elixir manufactured by the Standard Oil Company, remedied the constipation induced by sedentary urban life; Wrigley's chewing gum was a vast improvement on prehistoric days when people had sucked pebbles to keep their mouths moist; and the telephone was described as one of the means by which the 'civilised man is distinguished from the cave man'.[51] A striking example of how completely prehistoric imagery had insinuated itself into American commercial culture can be seen in a February 1917 edition of the *Seattle Star* newspaper. It is a two-page spread featuring twelve advertisements that used drawings of cave men and women to promote watches, flowers, pianos, dancing lessons, haberdashery, women's fashions, automobiles, apple juice and jewellery, along with a bowling alley and a builder.[52] Each advertisement described its particular product or service as an improvement over its archaic forebears.

Films were also emerging as the core of modern mass culture. British cinemas remained open throughout the war, while troops watched films in training, on leave and in periods of rest from front-line duty. Seymour

Hicks's *King Mugslot* was the first British prehistoric film released during the war, though American titles continued to be imported.[53] No early cave man film resonated as loudly as Charlie Chaplin's ten-minute feature *His Prehistoric Past*. It was released in the United States in December 1914 and in Britain in the spring of 1916, by which time Chaplin was the world's biggest film star.[54] He was also beloved by men in uniform because the Little Tramp's ability to overcome obstacles and outlast aggressive rivals suggested that they might survive the war. Chaplin recounted an anecdote about the film in his autobiography, saying, 'I started with one gag, which was my first entrance. I appeared dressed as a prehistoric man wearing a bearskin, and I scanned the landscape, I began pulling the hair from the bearskin to fill my pipe. This was enough of an idea to stimulate a prehistoric story, introducing love, rivalry, combat and chase.'[55] Though the film was made in California, where Chaplin had settled in 1913, its tone was essentially British. The plot about the prehistoric Little Tramp's rivalry for a young woman's affection with a larger Neanderthal was derived from Robey's 'Prehistoric man'. Chaplin also satirises D.W. Griffith's *Man's Genesis*, because the final scene sees a policemen wake the tramp, who has been dozing on a park bench. He is a downmarket everyman, in contrast to Griffith's protagonist who had fallen asleep in his private club.

The popularity of Chaplin's film, British stage plays with prehistoric scenes and cave man characters in advertisements helps to explain George Robey's decision to revive his fifteen-year-old 'Prehistoric man' sketch in January 1917.[56] It formed part of the revue *Zig Zag*, one of the war's most successful stage shows, which played over 600 times at the Hippodrome in Leicester Square. It was also a crowning peak in Robey's career. He had been an indefatigable war booster since 1914, raising money and men and serving as both a special constable and a firewatcher, roles that befitted his patriotic, conservative, middle-class persona. He had spent most of 1916 starring in the revue *The Bing Boys are Here*. In that production, Robey had played one of the titular brothers, overgrown schoolboys in short trousers and Eton collars, who had left their home in Binghamton for London. The light, escapist story had been designed to entertain troops on leave, many of whom may have recognised themselves in these innocents who navigated through the metropolis geographically and metaphorically. Robey's duet with the object of his affections, played by Violet Loraine, had been 'If you were the only girl in the world', which remains one of the war's best-known tunes.

Zig Zag was an altogether more exotic brew of short comic sketches with scenes featuring a chorus of ballet girls and grass-skirted prehistoric maidens designed to titillate men on leave. Robey first took the stage as the King of Neutralia, a bankrupt Ruritania that never went to war, a nod to the decisive roles that minor states had played in the current conflict. He then appeared successively as an apprentice barber and a naïve tourist from Accrington, Lancashire, in an echo of his *Bing Boys* role. But his most important contribution to the revue came when he donned the familiar red wig to play his prehistoric man, now known interchangeably as Mr Stone Hatchet or Algernon Umph, whose wife, played by the Australo-American Daphne Pollard, was the object of amorous advances from the young leading man.[57] The programme set the scene, for those not yet familiar with the sketch, by declaring that:

> Algernon Umph was a winsome boy
> The pride of the caves and the Stone Age joy
> Hector McHr-r-r! was a married gent
> Who lived in a bear-hole tenement
> And this is the tale of his tragic life
> With Pansy Maud, his flighty wife.[58]

The drama took place before an elaborate backdrop of prehistoric cave-dwellers, birds and beasts until, as before, Robey clubbed his rival and tossed his body into the sea.[59]

Uniformed men in the audience must have identified with Robey, given how often they had evoked such characters in trench newspapers and letters. At the same time, the plot about a husband dispensing rough justice to the man who was courting his wife provided a vicarious yet satisfying expression to anxieties about the fidelity of their girlfriends and spouses back home. Links between the sketch and soldiers' experiences became overt in September 1917 after Robey's fire-watching duties during a bombing raid left too little time to don his costume. So he performed the sketch in his lounge suit, wig and steel helmet. The war's intrusion into this fantasy inspired Robey to update the scene by having the backdrop repainted to show prehistoric soldiers living in caves to protect themselves from air raids carried out by flying dinosaurs. These new images tapped into pre-existing allusions about how Germany had reverted to a Neanderthal state and cartoons that artists in uniform had drawn, depicting the battle front as a prehistoric land derived from Reed's cartoons.[60]

Robey remained in London throughout the year, though a touring company took *Zig Zag* to provincial cities. In this production, a version of the

'Prehistoric man' sketch was enacted by the Egbert Brothers, second-rank performers best known for a sketch entitled 'The happy dustmen'. Robey's success also inspired comedians and creators of revues to incorporate prehistoric references. The least subtle appropriation may have been the Liverpool revue *Nothing New*, in which a historical pageant began with a prehistoric scene in which a cave man named George fought a love-rival. By the following autumn, Robey's entire sketch had been incorporated into a revue that toured Australia.[61]

Zig Zag resonated with servicemen from many countries, who made its song 'Over there' one of the war's best-remembered tunes. In mid-1917 a Canadian trench newspaper described Robey's performance as one of the delights of leave in London, while the following January another included a cartoon of Robey in his prehistoric costume in a gallery of British artistes who had entertained the troops. That July, a third Canadian newspaper lampooned an officer with a cartoon of him dressed as Robey's cave man.[62] Members of an Australian unit imitated Robey and Pollard in a concert, as did British prisoners of war in Germany.[63] When the Royal Flying Corps staged a sports day at Ascot in June 1917, it incorporated a sketch representing the history of technology that began with a cave man attempting to fly a prehistoric aeroplane.[64] It was a scene that reflected the place that prehistory had been awarded in Edwardian pageants.

Troops began giving prehistory new meanings in November 1918, when peace made the war's brutality and squalor seem like something from the earth's most ancient days. An American cartoon from the start of the year had foreshadowed the shift by depicting a cave man in German uniform, whose club drags dispiritedly on the ground. He is looking into the brilliant sunshine of the future as he reads the four principles for peace that President Wilson has just announced.[65] The image made it clear that the time for prehistoric barbarity had passed. In a similar vein, at year's end the Canadian Machine Gun Corps's trench newspaper reflected on the unit's wartime experiences in a series of articles entitled 'What they did: a prehistoric retrospect'. British units celebrated the Armistice with prehistoric revues that included fanciful dinosaurs and archaic hospitals, and officers wore cave men costumes to masqusrades, while the term was used to describe military boxers, and in March 1919 an American officer was criticised as a cave man for scrounging for souvenirs.[66] Finally, the *Athenaeum* magazine, which was not noted for humour, spoofed the connection between the war and prehistory in February 1920 by publishing the review of a book about Gloucestershire's ancient earthworks under the punning title 'Prehistoric man in the trenches'.[67]

Casual and commonplace references to prehistory also inspired the immensely popular Canadian humorist Stephen Leacock to write a short story in 1917 entitled 'The cave-man as he is', in which he declaimed:

> everybody nowadays knows all about the cave-man. The fifteen cent magazines and the new fiction have made him a familiar figure. A few years ago, it is true, nobody had ever heard of him. But lately, for some reason or other, there has been a run on the cave-man. No up-to-date story is complete without one or two references to him.[68]

Leacock appeased the narrative demands of his tale by giving the impression that cave men – a term he used in the American sense as referring to rugged frontiersmen – had arrived in a sudden onslaught of films, books and stage plays. This was not the case, since the character – in its broader, original British sense – had spread throughout the globe by the turn of the century. The cave man may not have appeared in a thunderclap, but wartime prehistoric allusions showed that people from throughout the English-speaking world shared a sense that prehistory was a comically archaic version of the present day. The imagery had provided those in uniform with a means of articulating wartime experiences, while also standardising the depiction of cave men that was rooted in E.T. Reed's cartoons.

Civilians had never lost their pre-war interest in prehistory. But as we will see, associating the ancient past with cosy assumptions about Britain's immemorial grandeur was increasingly incongruous after 1919. The country was economically shattered, the military was weakened, and so was the empire, since the Dominions and many colonies aspired openly for autonomy. The United States was emergent. Over the course of the twentieth century American popular culture would change cave men in ways that could not have been predicted.

Notes

1 'The evolution of the prehistoric pneumatic', *Western Daily Press* (22 April 1914), p. 9; Norman Kennedy, 'Prehistoric zig zags' (London: J.B. Cramer, 1913), British Library, Music Collections h.3826.oo(50); 'Interesting list of records', *Talking Machine World* (15 April 1914), p. 225.
2 'Penny rate for advertising', *Western Mail* (2 May 1914), p. 6.
3 'Local amusements', *Nottingham Evening Post* (26 May 1914), p. 3.

4 'Red resellers dance in a garden', *Daily Mirror* (19 June 1914) p. 8; 'Helping a hospital', *Daily Mirror* (1 July 1914), p. 8.
5 'Boy Scout notes', *Luton Times and Advertiser* (23 January 1914), p. 2.
6 'Prehistoric peeps at Bourneville', *Birmingham Gazette* (26 June 1914), p. 6.
7 The quotation is from 'Empire Theatre', *Coventry Evening Telegraph* (30 May 1914), p. 1. See also 'The operetta house', *Edinburgh Evening News* (28 July 1914), p. 4; 'The picture houses', *Birmingham Mail* (28 July 1914), p. 5; and 'Premier picture theatre', *Western Daily Press* (18 August 1914), p. 4.
8 'If the cave man visited the training camp', *Los Angeles Examiner* (26 June 1910), sporting section, p. 1.
9 'Hippodrome, Exeter', *Exeter and Plymouth Gazette* (23 July 1914), p. 1.
10 'Cinema drama at Portland', *Evening Despatch* (11 July 1914), p. 3; 'Seymour Hicks in prehistoric peeps – not strictly archaeological – at Portland: some very moving pictures', *Sketch* (29 July 1914), pp. 6–7; 'The King's Hall', *Dover Express* (26 March 1915), p. 8.
11 'In the Stone Age', *Colac Herald* (Australia) (9 January 1914), p. 1; 'Melba Theatre', *Argus* (Melbourne) (19 January 1914), p. 16; 'Tivoli Theatre', *Advertiser* (Adelaide) (2 March 1914), p. 19; 'New songs', *Marlborough Express* (New Zealand) (30 March 1914), p. 1; 'When man found fire – and light', *Newcastle Morning Herald* (Australia) (26 June 1914), p. 8; 'The theatres – Forty thieves', *Sun* (Canterbury, New Zealand) (3 August 1914), p. 9.
12 Winsor McCay (dir.), *Gertie the Dinosaur* (1914); 'Hammerstein's', *Variety* (27 February 1914), p. 19; '"Gertie" and other dinosaurs in the McCay picture', *Moving Picture World* (28 November 1914), p. 1242; Tom W. Hoffer, 'From comic strips to animation: some perspectives on Winsor McCay', *Journal of the University Film Association*, 28:2 (1976), p. 25.
13 Edgar Rice Burroughs, *The Eternal Lover* (Chicago: McLurg, 1925).
14 'We – and you', *Manchester Evening News* (26 September 1914), p. 3.
15 'Armageddon', *Manchester Evening News* (22 January 1915), p. 7.
16 The quotation is from 'Why Germany made the conflict', *Birmingham Gazette* (16 October 1914), p. 4. See also 'Joy of the khaki life', *Daily Mirror* (20 October 1915), p. 4.
17 'Grateful and comforting', *Daily Herald* (22 August 1914), p. 5.
18 'The outlook', *Birmingham Gazette* (21 January 1915), p. 4.
19 Will Dyson, 'Remember your illustrious forbears', *Daily Herald* (1 September 1914), p. 8.
20 Will Dyson, *Kultur Cartoons* (London: Stanley and Paul, 1915), unpaginated.
21 'Kultur cartoons', *Sun* (New York) (27 June 1915), p. 2.
22 'Substance and form: the guy that put the soul in soldier', *Life* (18 February 1915), Library of Congress, Swann no. 1585.
23 'Stone-Age tactics', *Day Book* (Chicago) (2 November 1916), p. 5.
24 See for instance 'Keep America out of war', *New York Tribune* (29 March 1917), p. 6.

25 W.A. Rogers, 'The gorilla of the sea', *New York Herald* (3 July 1918), p. 12, Library of Congress, CAI-Rogers, no. 238; Norman J. Lynd, 'Stop him!' (Philadelphia: Colonial Press, c. 1917), Library of Congress, 91796727.
26 See for instance 'The destruction of civilization', *Breckenridge News* (Kentucky) (11 April 1918), p. 5; *Durant Weekly News* (Oklahoma) (19 April 1918), p. 9; and *Exhibitors Herald* (20 April 1918), unpaginated.
27 Lewis Spence, 'The prehistoric Scout', *The Oxford Annual for Scouts* (Milford, c. 1918), http://www.trussel.com/prehist/spence.htm (last accessed 6 October 2016).
28 Ibid.
29 Lieutenant Thomas Treherne, 'An actor's war experience', *Era* (26 January 1916), p. 11.
30 'Footballers in army', *Lancashire Daily Post* (13 October 1917), p. 4.
31 'The heavies', *Graphic* (11 December 1915), unpaginated.
32 William Beach Thomas, 'Fantastic monsters', *Daily Mirror* (18 September 1916), p. 2.
33 'The new war monsters', *Chester Chronicle* (23 September 1916), p. 7.
34 '"Tank" starts for Berlin, but "juice" runs short', *Toronto Globe* (29 September 1916), p. 4.
35 The quotation is from 'Triangle Picture Hall', *Western Daily Press* (14 November 1916), p. 5. See also '"Juggernauts" Germans thought an impertinence: tanks in action', *Illustrated London News* (2 December 1916), p. 667; and 'Cornhill', *Aberdeen Journal* (3 February 1917), p. 2.
36 Malcolm Ross, 'The road to the front', *North Otago Times* (31 December 1917), p. 2; 'Extracts from the papers', *Chronicles of the New Zealand Expeditionary Force* (25 October 1918), p. 155.
37 'Bovington Camp', *Beaver* (22 February 1919), p. 6.
38 'How the tanks went over', *BEF Times* (1 December 1916), unpaginated. A similar spoof is 'We attack at dawn', *BEF Times* (15 August 1917), unpaginated.
39 The quotations are from, successively, J.A. Macpherson, 'Chronicles of maxim, no. 2: my first leave', *Canadian Machine Gunner* (December 1918), p. 11; 'To the soldiers of Canada', *Bramshott Souvenir Magazine* (1918), p. 44; Charles Sivell, 'Hunting beach chickens', *Canadian Machine Gunner* (November 1918), p. 16; 'War and national honour', *Dead Horse Corner Gazette* (October 1915), p. 2. See also 'Imaginary interviews with well-known men', *C.R.O. Bulletin* (24 July 1918), p. 3.
40 'Review column', *Trench Echo* (Easter 1916), p. 8.
41 'The war in the air', *(Last) Gasper* (30 September 1916), unpaginated.
42 See for instance 'Hospital evolution', *Searchlight* (October 1916), p. 3.
43 'Prehistoric pictures no. 1 – zero school of military aeronautics', *Gnome* (May 1917), p. 13; 'Prehistoric pictures no. 2 – an aeroplane co-operating with a tank', *Gnome* (August 1917), p. 14.
44 Hon. Mrs Dowdall, 'The boss woman', *Thistle: Scottish Women's Hospitals for Foreign Service Souvenir Book* (Glasgow: John Horn, 1916), p. 50.

168 Inventing the cave man

45 J.M. Collins, 'The caveman', *Anzac Book: Written and Illustrated in Gallipoli by Men of ANZAC* (London: Cassell, 1916), p. 113.
46 'Labour's loyalty', *Register* (Adelaide) (14 October 1915), p. 4; 'Labour's loyalty – ANZAC Day celebrations', *Observer* (Adelaide) (16 October 1915), p. 24.
47 See for instance 'Prehistoric Percy mobilises his troops!', *Burnley News* (27 February 1915), p. 4; 'Prehistoric Percy', *Burnley News* (10 April 1915), p. 9; 'Taking cover', *Burnley Express* (11 September 1915), p. 5; 'Prehistoric Percy slays the dragon', *Rochdale Observer* (16 October 1915), p. 5; 'Prehistoric Percy's bomb-proof shelter', *Burnley News* (6 November 1915), p. 8.
48 'The Panshine pair throughout the ages', *Birmingham Gazette* (27 October 1916), p. 6.
49 'Christmas fair by art students', *Glasgow Daily Record* (19 December 1916), p. 1; 'The coal shortage and the British fireplace', *Daily Mirror* (24 August 1918), p. 6.
50 'Athletic sports', *New Zealand Herald* (16 October 1915), p. 4; 'If you had lived as the cave man lived', *Townsville Daily Bulletin* (Australia) (23 November 1916), p. 7; 'For Easter holidays', *Advertiser* (Adelaide) (13 March 1917), p. 3; 'J.T. Calder', *Oamaru Mail* (14 August 1917), p. 5.
51 Rea Irvin, 'The Darwin club', *Life* (18 March 1915); T.S. Sullivan, 'Even in prehistoric times', *Judge* (26 February 1916); 'American Telephone and Telegraph Co.', *McLure's* (April 1916), p. 56; 'Nujol', *Topeka State Journal* (Kansas) (11 January 1917), p. 8; 'Wrigley's', *Puck* (3 March 1917), p. 29; 'He proves that Darwin was right', *Ogden Standard* (Utah) (12 October 1918), unpaginated.
52 'The cave man – Mr Stone Age', *Seattle Star* (13 February 1917), pp. 6–7.
53 See for instance 'Empire Theatre', *Coventry Herald* (26 March 1915), p. 1; 'Electra Palace', *Sheffield Evening Telegraph* (8 November 1916), p. 1.
54 'His prehistoric past', *Hull Daily Mail* (25 May 1916), p. 2.
55 Charles Chaplin, *My Autobiography* (London: Bodley Head, 1964), p. 164; Charles Chaplin (dir.) *His Prehistoric Past* (1914).
56 See for instance 'Dramatic and musical', *Liverpool Echo* (7 August 1915), p. 3; and 'Exchange Electric Theatre', *Taunton Courier and Western Advertiser* (23 May 1917), p. 4.
57 'Zig-Zag', *Illustrated Sporting and Dramatic News* (3 March 1917), pp. 14–15.
58 'A married gent, his flighty wife, and a winsome boy', *Sketch Supplement* (28 February 1917), p. 3.
59 A.E. Wilson, *Prime Minister of Mirth: The Biography of Sir George Robey* (London: Odhams, 1956), pp. 92–3; 'Zig-Zag!', *Stage* (28 June 1917), p. 10.
60 'The humorous side', *Hull Daily Mail* (26 September 1917), p. 2; 'The air raids', *Stage* (27 September 1917), p. 15; 'Fun and frolic in *Zig-Zag* at the Hippodrome', *Sphere* (3 November 1917), p. 110.
61 'Zig Zag', *Liverpool Daily Post* (5 April 1917), p. 3; 'Olympia', *Liverpool Daily Post* (10 April 1917), p. 3; 'With star "tops"', *Liverpool Echo* (27 April 1917), p. 3; 'Comedienne's "glamour" clothes for navy', *Hull Daily Mail* (10

January 1948), p. 1; 'Tivoli', *Liverpool Post* (19 March 1918), p. 3; 'New Empire', *Burnley Express* (23 March 1918), p. 2; 'HM Theatre', *Aberdeen Express* (30 March 1918), p. 2; 'Olympia, Stirling', *Stirling Observer* (20 April 1918), p. 4; 'Leith Alhambra', *Edinburgh Evening News* (30 April 1918), p. 2; 'Repertory Theatre', *Liverpool Post and Mercury* (26 December 1916), p. 3; 'Repertory Theatre', *Liverpool Echo* (26 December 1916), p. 3; 'At the Repertory Theatre', *Liverpool Echo* (12 January 1917), p. 4; 'As you like it', *Era* (31 January 1917), p. 15; 'In London' and '"Zig Zag" capital', both *Variety* (9 February 1917), p. 4; 'Amusements', *Age* (Melbourne) (28 October 1918), p. 9; 'In the limelight', *World's News* (Sydney) (21 December 1918), p. 5; 'Hello everybody', *World's News* (28 December 1918), p. 5.
62 L.J., 'What the stay-at-homes miss', *Tchun!* (30 June 1917), p. 7; 'All carrying on', *Canadian Machine Gunner* (January 1918), p. 31; 'The fossil king', *C.R.O. Bulletin* (24 July 1918), pp. 1 and 3.
63 'Military ball', *Mildura Cultivator* (Australia) (4 May 1918), p. 13; 'The British amateur dramatic society – programme – *A Jerry Built Genesis*', undated, Imperial War Museum, G. 2449-5.
64 'Royal Flying Corps sports on the racecourse', *Reading Mercury* (2 June 1917), p. 6.
65 'Blinding light', *Dayton Daily News* (Ohio) (13 February 1918), Library of Congress, Swann no. 1670.
66 'What they did: a prehistoric retrospect', *Canadian Machine Gunner* (November 1918), pp. 8–9; 'The acks' concert', *Bedfordshire Times* (27 December 1918), p. 5; 'Nevers bouts draw big holiday crowd', *Stars and Stripes* (10 January 1919), p. 6; 'Dance notes', *C.R.O. Bulletin* (15 February 1919), p. 2; 'Bullet hunt is all the rage at Coblenz', *Stars and Stripes* (7 March 1919), p. 7; 'In Bonn', *Stage* (31 March 1919), p. 15.
67 'Prehistoric man in the trenches', *Athenaeum* (6 February 1920), pp. 176–7.
68 Stephen Leacock, *Frenzied Fiction* (London: Bodley Head, 1917), p. 113.

7

Modern times: the Victorian cave man's long afterlife

In November 1919, the novelist H.G. Wells began publishing a global history in fortnightly instalments that was compiled the next year as *The Outline of History*, a book that eventually sold over two million copies. History was a new field for Wells, so he worked with a team of experts, including the Natural History Museum's Ray Lankester, who advised on the ancient past. Wells created an accessible tone by larding his text with illustrations, chatty asides and footnotes, including one in the chapter about dinosaurs explaining that:

> the genius of a great humorous artist (E.T. Reed) obliges us to add a footnote to clear away a common misconception. He was the creator of a series of fantastic pictures, *Prehistoric Peeps*, which have had a deserved and immense vogue, and it was his whim to represent primitive men as engaged in an unending struggle with great Plesiosaurus and the like. His fantasy has become a common belief …[1]

Wells then stated the obvious fact that dinosaurs and cave men had never coexisted. Reed responded to the aside good-naturedly with a cartoon, which was published in the *Bystander*, of Wells and Lankester huddling in stammering fright as they are stalked by a fanciful dinosaur (see Figure 22).[2] Wells was among the intellectuals who believed that the war had ruptured civilisation, ushering in an age that the historian Richard Overy characterised as 'morbid' because of a pervasive sense of impending crisis.[3] But Reed's cartoon, which was rooted in Victorian ideals, suggested that popular culture had survived and perhaps even been strengthened by the conflict. The image is a nice introductory metaphor for how comic prehistory would continue to cow attempts to create a countervailing and scientifically accurate popular understanding of the ancient past.

Evidence of comedy's hold on British conceptions of the ancient past

"Who Says I'm 'Not Contemporary With Man'?"

SIR RAY LANKESTER: "I d-do b-b-believe that awful B-B-Bronto-Saur'us!"
H. G. WELLS: "Who, Ray? Who, Ray? (I d-d-don't m-mean 'Hooray') W-why, you told me '*P-P-Prehistoric Peeps*' w-were all wrong!" *(See footnote to "Outline of History" Part 1)*
BY E. T. REED

H.G. Wells and Ray Lankester menaced by a dinosaur, *Bystander* (24 December 1919).

is widespread. In November 1919, Stanworth's Umbrellas revived the cartoon cave man Prehistoric Percy in advertisements that appeared until at least 1925. Stage evocations included the scene in the 1919 Christmas pantomime at London's St Martin's Theatre in which club-wielding cave men wooed a young girl before being devoured by a dinosaur. The following year, pupils at Bedford School produced a spoof newspaper entitled the *Prehistoric Times*, whose cover featured a cartoon of a woman milking a dinosaur.[4] Newspapers also show that fancy dress balls, summer fetes, carnivals and pageants from London's West End to Whitby, Hull, Barnstaple and Bedford regularly included *papier maché* dinosaurs and people dressed as cave men.[5] The most notable of them was the week-long spectacle that over 2,000 actors staged at Liverpool's Wavertree Park in September 1930 to mark the centenary of the Liverpool and Manchester Railway, which had operated the world's first passenger train. Fittingly, the celebrations began with a 'historic pageant of transport' designed to show how this British invention had been exported around the world and had largely defined the nineteenth century. The event adopted the structure that had become commonplace in the Edwardian era by beginning in Britain's prehistory, brought to life in this case by a club-wielding cave man who dragged a woman by the hair as they were pursued by a fanciful dinosaur.[6]

The year before, a rugby team clad in fur and armed with stone axes had enacted 'prehistoric football thrills' while being chased by a dinosaur at a Somerset carnival.[7] Such matches had become so commonplace by 1937 that the organisers of a Gloucester summer fete decided to stage their pageant in reverse. It opened with a display of aeroplanes and cars and regressed in time through Henry VIII and Robin Hood, before concluding with a prehistoric rugby match that was interrupted by a dinosaur.[8] Rugby was a particularly rough, though middle-class, game that lent itself to prehistoric comparisons in much the same way as other sports had done before the war.

Such prehistoric scenes reinforced the teleological belief that national ideals and ingenuity were rooted in the long-ago past. Prehistory may have been chaotic and violent, but positioning it as the foundation for the invention of the steam locomotive, the symbol of Victorian industrial might, or as the roots of rugby, a middle-class public school game, suggested that the British had channelled cave man energies into ever-upward progress. It was a comforting bulwark of the war-weakened belief in the nation's eternal greatness.

An equivalent desire to root national pride, confidence and codes of gentlemanly behaviour in the country's ancient past lay behind prehistory's appeal for the Scouting movement. Its founder, retired General Sir Robert Baden-Powell, had skilfully promoted his ideas from the first camp at Brownsea Island in 1907 through the wartime addition of Wolf Cubs for younger boys. The cubs' *Handbook* drew heavily from *The Jungle Book*, Rudyard Kipling's tale about an English boy raised by animals in the Indian forests. The latter was not a cave man story *per se*, but as with *Tarzan* of 1912, there were obvious analogies in these tales of English boys raised in the wild. Scouting was a largely decentralised movement in which leaders set the tone for their local troops. Even before the war, some of them saw re-enacting prehistoric scenes as an amusing way of fostering woodcraft, stalking and hunting skills (see Figure 23). As a Scout from Battersea fondly recalled, his leader, the local curate, regularly clad the boys in costume to stage mock battles between ancient Britons and savage invaders.[9] The Britons' inevitable victory proclaimed national superiority.

Such local scenes were echoed in Scouting's national and international meetings, known as jamborees, which included historical parades consciously modelled on pre-war pageants.[10] So it is not terribly surprising

MR. ATWELL — FOR HIS BONY PART AS A PREHISTORIC MAN AND FOR WALKING BIRMINGHAM STREETS AS HERE SHOWN.

A prehistoric Scout leader, *Sketch* (9 July 1913).

that at the first international jamboree, in London in 1920, British Scouts evoked the nation's history by demonstrating Morris dancing and equally by building an enormous Reed-inspired comic dinosaur, christened the 'Onestaruorus', that they manoeuvred past Baden-Powell as he reviewed their parade.[11] At Glasgow three months later, Field-Marshal Sir Douglas Haig, who had commanded the British Expeditionary Force in the First World War, saw Scouts parade a fanciful dinosaur named the 'Jamborivisaurus'.[12] The boys who made and animated the dinosaurs were inspired by civic pageantry, while the Morris dancers reflected the efforts of modern folklorists to recapture what they believed to be a fundamental, pre-industrial British spirit. More research is needed to fully understand the role of prehistory in the Scouting movement, but references over the coming years indicate that Stone Age cricket and football matches, mock combats and domestic scenes in local and national pageants were fairly commonly staged by Cubs and Scouts until at least the early 1930s.[13] The phenomenon followed the movement as it spread across the globe, with Australian and New Zealand Scouts building giant dinosaurs and staging prehistoric sports for their own jamborees.[14] In those two countries, cave men also continued to be invoked in speeches, revues and pageants, and were depicted commonly in advertisements for both local businesses and American-owned products like Wrigley's chewing gum (see Figure 24).[15]

Despite these examples of resilient British interest, the post-war story of cave men is largely tied to the emergence of the United States as the global cultural capital. American Boy Scouts staged prehistoric sports, while cartoons, advertisements, plays and films featured cave men, though the character continued to have a slightly different meaning in the United States.[16] For instance, the popular 1922 self-help book *The Caveman Within Us* encouraged Americans to harness their rough inner masculinity, which as we have seen, embodied the national frontiersman ideal.[17] A more negative concept of masculinity was articulated that November, when the *New York Times* reported that three men were petitioning the state Supreme Court to issue a charter for the Association of Brothers Under the Skin, whose purpose was to reclaim men's unchallenged authority by among other things developing '"cave man" methods for the better discipline of jealous, nagging or unreasonable wives'.[18] The request was simply a publicity stunt to promote a film of the same name. But the credulity with which the American press reported the story exposed the way in which the term 'cave man' had been used for several years in newspapers and magazines

Prehistoric chewing gum, Age (10 October 1927).

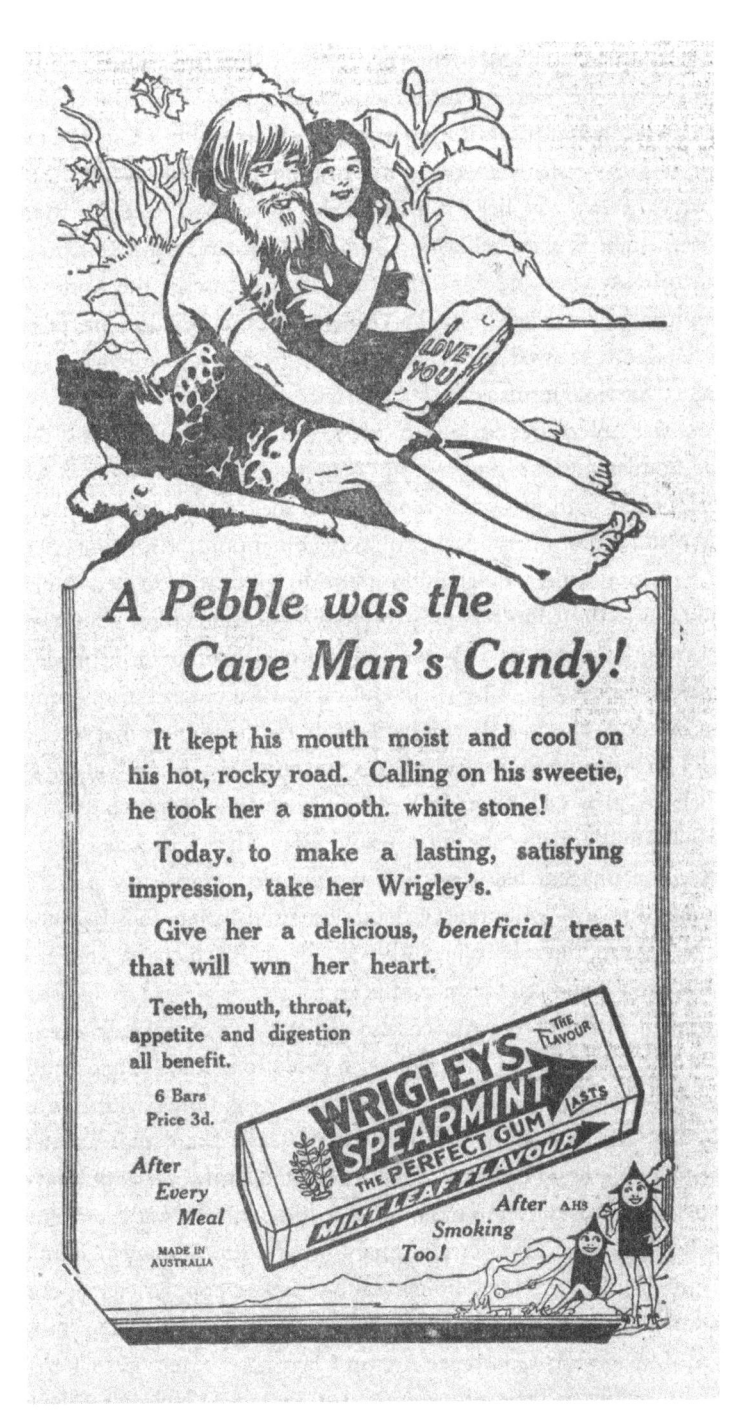

as a light-hearted and, by twenty-first-century standards, shockingly inappropriate way of referring to domestic abuse. Stage and film depictions were one thing, but as a Los Angeles police officer would say in 1924 while attempting to persuade a woman to prosecute her abusive lover, 'the cave man style of love was illegal outside of the Hollywood screen'.[19] Men did not learn such abusive behaviour from the courtship rituals portrayed in prehistoric sketches and films, but the frequent use of the term, which seems to have travelled from the United States to Britain, Australia and New Zealand, showed that such scenes provided a dismissive frame of reference for violent misogyny.[20]

Post-war American allusions to cave men also took place against the rise of the Fundamentalist movement of conservative Protestants who refused to adapt their faith to secular thought and society. Fundamentalism suddenly made evolutionary ideas extremely contentious, surprising scientists who had long found an equilibrium with faith, and divided American Protestants – the country's most numerous denomination – into mutually antagonistic camps. This made individuals who could reconcile faith and science 'almost invisible'.[21] Evolution was denounced from pulpits, at camp meetings and in a slew of periodicals and pamphlets that were often heavily illustrated with images of apes and monkeys. Many fundamentalists believed that science, and Darwinian theories in particular, had been used immorally during the First World War. Their cartoons often drew on wartime imagery like depicting the populist, three-time presidential candidate and anti-evolutionary champion William Jennings Bryan as the hero of Verdun, the massively costly 1916 battle in which the French army had withstood repeated German attacks.[22]

Fundamentalism's emergence also shows how completely cave men had been divorced from direct links to evolution, since religious ire was focussed on scientists and such authoritative evolutionary displays as the Hall of the Age of Man, which opened in 1921 at Manhattan's American Museum of Natural History. The hall's glass and hardwood exhibition cases held an extensive collection of hominid fossils, while its walls were covered by paintings depicting an evolutionary progression of ancient hominids. The murals were extremely influential in shaping popular conceptions of prehistory in the United States for decades, demonstrating the immense power of images. As Constance Areson Clark has written, in that country 'no matter how many words evolutionists wrote acknowledging the complexity of evolutionary patterns, the public discourse was saturated with visual allusions to a linear, goal-directed and hierarchical version of evolution.'[23] This was evident in the 1921 animated film *The First Circus*, whose introductory title card read 'Forty years ago today, Charles Darwin wrote his book "Descent of Man from Monkey"', a deliberate misattribution that signalled the film's anti-evolutionary message. Knuckle-walking hominids are seen discovering and consuming a bottle of alcohol, a comment on unevolved instincts that could be cured through Prohibition, which had been introduced in 1920. The film concludes with a performance by the simian acrobats of the 'Stonehenge circus', in a scene that further attributes corrosive evolutionary ideas to Britain.[24]

The American battle between science and faith was most famously waged in the summer of 1925, when John Scopes was charged with teaching evolution in a Tennessee high school, which was illegal in state-funded institutions. The eight-day trial captivated the nation, becoming a showdown between Bryan and the atheist Chicago lawyer Clarence Darrow. Both sides broadcast their beliefs through cartoons of apes, many books and at least one film purporting to trace human evolution. The result was never in doubt. After deliberating for only nine minutes, the jury convicted Scopes. The verdict was soon overturned, though there was no retrial. Instead, Fundamentalists retreated into Christian colleges before re-emerging in the 1950s.[25]

A less intransigent Christianity is seen in a 1925 passage from the British Roman Catholic apologist G.K. Chesterton, who differentiated between the concept of evolution and the cave man in popular culture. Chesterton accepted scientific evidence about ancient hominids without believing they were related to modern humans. At the same time, he noted sardonically that:

> to-day all our novels and newspapers will be found swarming with numberless allusions to a popular character called a Cave-Man. .... So far as I can understand, his chief occupation in life was knocking his wife about, or treating women in general with what is, I believe, known in the world of the film as 'rough stuff.' I have never happened to come upon the evidence for this idea.[26]

To Chesterton, the cave man had been invented by cartoonists and filmmakers, not scientists. He was no fan of the character, but his faith was not threatened by it.

Chesterton could well have been inspired by two important films that appeared immediately before he wrote this passage. Modern cave man films and the full adoption of this British character in American popular culture date from Buster Keaton's 1923 feature *Three Ages*. Keaton, who had grown up on vaudeville stages, was emerging as one of the greatest film stars. His comedy was founded on incredible athletic ability, while his screen persona has been described as a 'stony-featured clown, ethnically and socially marked as an Anglo-American working man of the Midwest' who was transformed in film after film into a hero.[27] In other words, his utter Americanness made him the nation's perfect cave man.

With *Three Ages*, Keaton forsook the twenty-minute short films that had been early cinema's staple for the hour-long features that were its future. The plot stitches together stories depicting love in the prehistoric, Roman and modern ages because, as Father Time intones in the first scene, 'Love is the unchanging axis on which the world revolves.' The Stone Age scenario is a love triangle in which Keaton vies with a larger, hirsute and more brutish rival for the affections of a fair-haired girl. Keaton is clad like one of Reed's cave men, rides a brontosaurus in a sophisticated bit of stop-motion animation and inhabits an archaic version of the modern world, in which a Ouija board satirises the post-war vogue for the occult, while he golfs – a gag directly stolen from George Robey – and visits a chisel-wielding stenographer. The audience see Keaton perform many dangerous stunts before knocking his rival off a cliff and dragging a beaming woman away by the hair. The music scores distributed with the film, which cinema organists would play to accompany the action on screen, underlined that this was an anachronistic vision of contemporary society by including current songs like 'Running wild' and 'Toot toot tootsie'.[28]

Scholars have generally viewed *Three Ages* as a send-up of D.W. Griffith's films *Man's Genesis* and *Brute Force*, or tied it to a sub-genre of early American cinema that expressed moral lessons about the contemporary world by juxtaposing modern plots with scenes set in ancient antiquity.[29] These are sensible epistemological views, though the important relationship between *Three Ages* and the British public has been relatively overlooked. The film was released in Britain in June 1923, three months before it opened in the United States, with a benefit screening attended by the dowager Queen Alexandra and Princess Alice. The event launched a marketing campaign built around Margaret Leahy, who played the object of Keaton's affections. She was a London shop assistant who had ended up in California after winning a *Daily Sketch* newspaper contest to transform an ordinary young woman into an American film star. Leahy had not been intended to play a cave woman, but when the studio had realised that her physical attractions were unmatched by talent, she had been fired from one film and assigned to Keaton, on the pretext, perhaps, that a prehistoric role would not be taxing. But Keaton and his crew were soon deeply frustrated at having to shoot the simplest scenes time and again. *Three Ages* was both a considerable financial success and Leahy's only film.[30]

The heavy marketing of *Three Ages* in Britain capitalised on Leahy's story and the long-standing public interest in prehistory. British audiences must have recognised how closely Keaton's costumes, gags and plot echoed Robey's 'Prehistoric man' sketch. Robey had starred in long-running wartime revues and been invested as a Commander of the Order of the British Empire for raising troops and money, but his future was uncertain. At fifty-four he no longer possessed an athlete's build, while cinema stardom had eluded him. His first film had appeared just before the war, and *George Robey's Day Off* of 1919 had showed him gardening and performing at a charity concert on a 'typical' rest day.[31] It had failed, despite the backing of Sir Oswald Stoll, who owned Britain's most extensive syndicate of variety theatres and was investing heavily in cinemas. Robey had made four equally unremarkable films for Stoll in 1923.[32]

When *Three Ages* reignited interest in Robey's sketch, Stoll financed the filming of an updated version in Devon in November 1923. The result, which was released the following April, married a dismal imitation of Keaton's physicality with an unfunny transposition of Robey's sketch to the silent screen. The scenes in the print in the British Film Institute are out of sequence, a common feature of surviving early films, but they can be easily reordered.[33] The opening title card mimics Robey's stage persona by declaring, 'Our story concerns the Pre-Historic Age when all the world was young – when men were men – magnificent – robust.' Robey's character has been renamed 'He of the beetle brows' in reference to his trademark eyebrows, while his love interest, played by Marie Blanche, has been rechristened 'She of the permanent wave', an up-to-date allusion to the hairstyle favoured by the young jazz set.

Robey's entrance astride an eight-legged pantomime dinosaur, using a massive club for a crop, is another nod to Keaton and to the dinosaurs that proliferated in British pageants. When Robey learns that his sweetheart is being put up for auction, he charges off, and the film cuts to a scene that is introduced by an intertitle reading 'pre-historic auction room about the year one million AM (Probably earlier, editor)'. A rapid series of intertitles then attempts to convey the frantic activity in the sales room. Robey's love-rival declares, 'I bid my priceless collection of cigarette cards of famous billiard markers', to which he replies 'that's my cue', and a woman says 'don't let him baulk you', while the auctioneer asks, 'Any advance? Are the rest of you all snookered?' After a card declares that the 'Beetle makes a break all around the table', Robey bops over a dozen opponents with his enormous club, before 'Game!' flashes on the screen as he hits the final one. This recapitulation fairly accurately recreates the film's tone-deaf attempt to transfer Robey's stage act to the screen, if an auditory metaphor can be used to describe a silent film. The frantic scene requires the intertitles to be spoken with timing and intonations that took years for performers like Robey to master. They could never be read as adeptly by an audience and so they interrupt the action.

The rest of the film is a jumble of satirical commentaries on modern Britain. Robey drives a prehistoric car with wooden wheels. A shot of the exterior of a Lyons Popular Café, one of the ubiquitous high street tea shops, dissolves punningly into 'Lyons pre-historic – Poplar', referring to a raucous place where Robey and his girl dance the jazzy 'Prehistoric prowl', evoking earlier cave man songs in audiences' minds. When she refuses to marry him 'until the Stonehenge Wanderers have won back the cup from the Rock Ramblers', a football match begins that is indistinguishable from a riot, or the prehistoric sports played at carnivals throughout the country.[34] Audiences are meant to laugh at an intertitle reading 'Result – Only fifteen dead or dying – apart from that it was only a friendly game.' The scene in which Robey seeks the consent of the girl's parents is a classic music hall take on working-class aspirations. Her father questions Robey's ability to support a wife, agreeing only after he declares, 'I am on the dole!', which is greeted with open-armed joy. A shot of the exterior of Selfridge's department store, a joke taken directly from Seymour Hicks's 1914 film *King Mugslot*, dissolves to a cave-like interior with an archaic lift and racks of clubs for sale. Robey then learns that his sweetheart's house has caught fire thanks to the heated language exchanged between her father and another suitor. This prompts 'The charge of the Fire Brigade', in which a helmeted Robey declaims through an intertitle card, 'Yours is not to ask the reason why / Yours is but to do a guy.' The fire brigade arrives to discover an old man playing patience. Robey asks about the fire, only to be told that it is 'In the grate', an age-old pun that causes everyone to collapse as the film ends.[35]

The film was a financial disappointment and unsurprisingly, Robey's career was largely eclipsed as live variety entertainment gave way to cinema.

The failure also dissuaded British filmmakers from exploring prehistoric subjects for more than forty years. Cave men settled in Hollywood, where Keaton's success had finally implanted the British idea that prehistory was an inherently comic era that could be used to satirise the modern world. It was also a convenient way for American filmmakers to re-envision prehistory, because films that claimed to depict the ancient past with scientific accuracy, as many pre-war ones had done, might now attract Fundamentalists' wrath. So, with few notable exceptions, American cave man films ceased making sober social commentaries. The shift, incremental and implicit as it was, had a profound effect, because it set cave man comedies on a gradual downward trajectory into very cheap productions, short farces and cartoons. The impact had global repercussions because throughout the twentieth century, American studios distributed their output very widely in overseas markets like Britain to increase profits.

None of this was pre-ordained in 1928 when Hal Roach shot the seventeen-minute cave man comedy *Flying Elephants* in the Nevada desert. Roach attempted to capture the spirit and look of Keaton's *Three Ages*, but the unhappy studio did not release the film until a year later, by which time two of its cast, Stan Laurel and Oliver Hardy, had become big stars.[36] True to most previous cave man comedies, the adipose Hardy played a swaggering club-carrying giant, who competes for a woman's affections with Laurel's bashful, flower-picking youth. The opening intertitle reads '6000 years ago all men were forced to marry or work on the rock pile – That's why it was called the Stone Age.' The film is a collage of sight gags about an archaic version of the modern world in which cave men ride cows, club one another and train for marriage by wrestling a real bear, while a bullock cart with solid wood wheels is called 'Ye firste Ford', a gag about cars that was now almost as ancient as the prehistoric conceit. The cave women are dressed like flappers, the archetypal 1920s party girls.[37] *Flying Elephants* is not very funny and made little money, dissuading American filmmakers from returning to prehistory for over a decade.

Film studios may have lost interest, but cartoonists continued exploring the possibilities of prehistory. E.T. Reed died in 1933, while Lawson Wood, Frederick Burr Opper and his other imitators had long since moved on to new subject matter. A younger generation began publishing in magazines like the *New Yorker*, whose cartoons are the best modern embodiment of *Punch*'s humour. Dozens of cave man cartoons have appeared in the magazine since at least the 1930s.[38] Cartoonists who drew newspaper comic strips set in prehistory included V.T. Hamlin, who first drew *Alley Oop*, about a time-travelling Neanderthal whose adventures satirise twentieth-century American life, in 1932.[39] The comic strip is still widely carried by newspapers, as is Johnny Hart's *B.C.*, a series about cave men and fanciful dinosaurs that that began in 1958 and plays in much the same slapstick register as Opper. American film studios have also produced many prehistoric cartoons, beginning with *Daffy Duck and the Dinosaur*, an eight-minute 1939 short that, so the prologue explained, presented the story of Casper Caveman hunting a prehistoric Daffy Duck 'millions and billions and trillions of years ago'.[40] It was a mildly funny diversion that lacked any profound message.

Hal Roach evidently still believed in the era's cinematic possibilities, because he returned to it in 1940 to make *One Million BC*, one of the most important, influential and financially successful prehistoric films of all time. Roach eschewed comedy for the moral, though not scientific, lessons imparted by early American cave man films. In the opening scene, hikers shelter from a storm in a cave. They discover a scientist deciphering wall drawings, who tells them that a 'learned man left a saga' about an ancient people who felt the same emotions, thoughts and struggles as we do, imbuing the film with a fatuous intellectual credibility and identifying the subjects of this ancient tale as the audience's prehistoric analogues. It is also a topical commentary on the modern world. The hikers' Tyrolean hats and lederhosen suggest middle Europe, while the story involves Tumak, a hunter from the Rock People, a dark-haired tribe that represent the Nazis, being described as cruel hunters who had to 'kill for sustenance; they despised weakness, worshipped strength, ruled by the power of might. The strongest was their leader.' Tumak loves Loana from the fair-haired River People, whose bucolic, egalitarian, existence represents democracy. The scientist's lecture dissolves to scenes of Tumak and Loana playing out their romance until a cataclysm destroys their world. In the last scene, the pair walk into the sunset accompanied by a child they have rescued. It was an unsubtle message.[41]

One Million BC was a huge box office success, but it was not followed by other prehistoric features, perhaps in part because the studios turned to more straightforwardly patriotic fare as the war loomed. Instead, the film inspired the 1940 cartoon short *Prehistoric Porky*, in which Warner Brothers' stammering pig and his pet dinosaur search for a suit of clothes, and *The Art of Self Defence* of 1941, in which the cartoon dog Goofy explores the sport of boxing from the Stone Age to the present.[42] A long downward slide began with *The Ape Man*, of 1943, in which Bela Lugosi, the Hungarian horror icon, played a scientist who injects himself with a secret formula and, predictably, turns into a proto-human simian. It was followed a year later, equally predictably, by *Return of the Ape Man*.[43] The war's end brought little immediate change of tone. *Prehysterical Man* of 1948 follows 'Popeye's anthropological expedition' to Yellowstone Park, where a Neanderthal clubs Olive Oyl and drags her away by the hair. Popeye eventually eats his spinach, rescues Olive and punches a brontosaurus with such force that only a skeleton remains, on which is affixed a sign reading 'Dinosaur, now extinct'.[44] That year also saw *I'm a Monkey's Uncle*, a short film starring the Three Stooges, the vaudevillians famed for extremely violent comedy. The written prologue states punningly that 'according to Darwin our Ancestors hung from trees by their tales – and thereby hangs our tale'. But the film has nothing to do with science. Instead, the prehistorically clad Stooges employ archaic technology and courtship rituals before fighting their rivals in a scene that recycles age-old gags. An extended version of the film was released in 1955.[45]

Some of these cheaply produced films aimed to comment on the contemporary world, an objective that was much more fully realised in the 1958 cartoon *Prehysterical Hare*, whose title suggested it had been inspired by the earlier Popeye cartoon. In the opening sequence the lisping hunter Elmer Fudd chases Bugs Bunny into a cave whose walls are covered in an ancient script that is translated to read 'Time capsule Circa 10,000 B.C. To be opened 1960 A.D.' It contains a reel of film that Bugs plays on his home projector, one of the signifiers of American middle-class prosperity. The title card equally proclaims post-war comfort by announcing that the film was shot in 'Cromagnonscope' with 'Neanderthal Color', allusions to the era's many vivid film processes. But the script was not complacent, since a basso voiceover proclaims that the film 'has been made to preserve a record of our way of life, since we know that other civilisations have perished', echoing a fear that many had for the United States during the Cold War. What follows is a fairly typical Bugs Bunny cartoon projected back into prehistory, in which the renamed Elmer Fuddstone stalks what he calls a 'sabre-tooth wabbit'.[46]

Prehysterical Hare demonstrated that prehistory could be used to satirise the modern United States, much as E.T. Reed had done to Britain at its imperial zenith. The United States was the most powerful nation on earth, just as Britain had been in Reed's era. For white Americans, it was the age of life-long jobs, chrome-bedecked automobiles, sprawling suburbs, swimming pools, television and teenagers. But it was not all optimism. Blacks struggled for civil rights, the Cold War threatened apocalypse, and, more prosaically, the American film industry was in crisis. In May 1948, the Supreme Court had ruled that studios' ownership of cinema chains amounted to illegal monopolies. Traditionally, Hollywood studios had produced inexpensive, reliably profitable, relatively short B-movies – including virtually all the examples cited above – which they had paired with more lavish 'feature' films in cinemas that they owned. The court had struck down this business model because it had denied competitors fair access to exhibition spaces. Studios had reacted by concentrating on lavish, vividly colourful blockbusters – many of which were epic versions of Biblical and Classical stories. These played for months in downtown cinemas.

The legislated industrial realignment coincided with a dramatic drop in cinema attendance, due in part to the advent of television and the growth of modern suburbs, since cinemas were traditionally concentrated in urban cores. Some of the decline was offset by the almost 3,000 drive-in cinemas that were built between 1946 and 1953. The drive-in was one of the most potent symbols of post-war American car culture. It was based on a business model that paired recent feature films with ones produced specifically for the new market, and in the process spawned a genre of American cinema. Teenagers were the biggest segment of the drive-in audience, since many had access to family cars and were especially keen on movies during their summer holidays, a traditionally fallow period in the pre-air-conditioning era. Teenagers wanted action, sexual excitement and young stars. Producers like Sam Arkoff understood this new audience and, as he put it in an unabashed preference for profit over art, 'advertised our pictures for the young dating crowd. We found our niche and stuck with it.'[47] Arkoff was one of the new chronically under-capitalised independent filmmakers who pared every extra penny by making movies in as little as two days, having actors play multiple roles, shooting on existing sets and outdoor locations that required minimal lighting, and using inexpensive black and white film stock. It was a canny strategy, because films that cost so little were reliably profitable, while those that caught the public imagination generated fortunes. Cave man films, which require almost nothing in the way of costumes or sets and give ample opportunities for action and the display of young bodies, have been a staple of low-budget filmmaking ever since.[48]

The impact of the new approach and new audience was most notable in the portrayal of cave women. They had been sexualised for decades as objects of male lust on stage and in films, being clad as provocatively as possible, given contemporary social mores. They had also been sexualised far more overtly in images created for private consumption, like the stereoscope cards of bare-breasted cave women that had been sold in France at the turn of the century, or the American 'cave dancer' Ethel Wilson, who toured the country during the First World War with a show in which she declared provocatively that her 'wardrobe mistress is Mother Nature'.[49] The trend was amplified in mid-century as mainstream productions began portraying cave women as libidinous, buxom, bouffanted and bikini-clad. This remarkably persistent voyeuristic fantasy was evident in the 1950 film *Prehistoric Women*, whose pseudo-scientific prologue proclaimed that it was the 10,000-year-old story of Tigri, the leader of a female tribe who dance orgasmically under the full moon. They capture men from a less evolved tribe and fight one another to determine with whom they will mate.[50]

An equally emergent adolescent vision of prehistory is best captured by *Teenage Caveman*, the 1958 film by Roger Corman, the greatest bad filmmaker of all time. Corman shot it in an abandoned quarry in ten days for $70,000 under the title *Prehistoric World*. Arkoff felt that this sounded like a boring museum exhibition, so he insisted on a more lurid name that tied it to his very profitable recent productions *I was a Teenage Werewolf* and *I was a Teenage Frankenstein*.[51] Robert Vaughn plays a Brylcreemed prehistoric youth who bristles at the ancient laws forbidding his tribe of bare-chested hunters and elaborately coiffed but scantily clad women from crossing a sacred river to a land where a deadly creature is said to live. Much ersatz philosophising and ponderous declarations like 'the law is old and age is not always truth' – this young man is hardly an angst-ridden James Dean – occur before a group led by Vaughn fords the river and kills the monster, which turns out to be an astronaut. His possessions reveal that a nuclear war has plunged humans back into a prehistoric existence. In the final scene Vaughn leads a party in search of new lands as a voice declaims portentously, 'this cycle has happened before. Will it happen again and if it does, will anyone survive?'[52]

Teenage Caveman is a masterpiece in comparison to *Eegah* of 1962, an extremely inexpensive collage of dune-buggying, droning rock music and southern California suburbia. Eegah, played by Richard Kiel, best remembered as the villain Jaws in two James Bond films, is a Cro-Magnon who has survived in the desert only to be discovered by a teenager as she drives to her country club. She and her father, a wealthy businessman with a convenient interest in palaeoanthropology, are eventually imprisoned in Eegah's cave, whose walls are covered in paintings. She remarks that they look like the ones 'in the cave in France' in a commonplace attempt to inject scientific credibility into a deeply frivolous production. When the pair escape, Eegah pursues them through the affluent city, before being cornered and shot by the police at a backyard pool party. The film closes with the father declaring, 'Yes he was real, it says so in the Book of Genesis. There were giants in the earth in those days', another leaden attempt to endow the film with meaning.[53]

As we have seen, British filmmakers and stage performers largely lost interest in prehistory before the Second World War, even if the public continued to recreate the era in pageants and sports events. Two references from 1949, inspired by preparations for the Festival of Britain, the showpiece intended to rekindle national pride, demonstrate that comic prehistory still resonated with people of a certain class and generation. Firstly, the cartoonist Osbert Lancaster spoofed antiquarian pedantry in an account of the ancient town of Draynefleet, whose name suggested both sewerage and sleekness. He passed over the settlement's earliest era by saying, 'it is conceivable, though admittedly there is little evidence to suggest it, that primitive man dwelt here before even there was a river at all, at a time when France and England were joined by a land-bridge and vast mammoths and sabre-toothed tigers prowled through the tropical undergrowth where now stands Marks and Spencers'.[54] The passage tied the most resonant symbol of modern British consumer culture to an age-old past. That July, Wilson Harris, the sixty-six-year-old MP for Cambridge University, stood in the Commons and asked whether the responsible minister would 'do his best to prevent any honourable member of the House from appearing in the "prehistoric peeps" section' of the festival.[55] The quip cast the event as an outdated attempt to recapture Britain's glory, while also recalling the pageants and parades in which prehistoric scenes had helped to project a sense of immutable British greatness. The men themselves seemed to be relics of an earlier age. Harris was among the last university representatives at Westminster, as the seats, which privileged social elites, were abolished in 1950. Though Lancaster was a generation younger, his luxuriant moustaches, Mayfair persona and affection for British eccentricities belonged to Reed's era. A more serious, if no less populist exploration of the nation's ancient past was also being reframed in popular books like *Looking at History: Britain from Cavemen to the Present Day*, a heavily illustrated 1955 volume that eventually sold eight million copies.[56]

Humour was resilient and reasserted itself in 1960, when the Oxford-educated economist Roy Lewis published *What We Did to Father*, a Stone Age satire on modern Britain. The narrator's social aspirations make the book a darker, prehistoric version of *Diary of a Nobody*, George and Weedon Grossmith's 1892 satire of lower-middle-class life. This articulate, red-haired youth named Ernest has ambitions of moving into a more commodious and better-located cave. His father is eager for the tribe to adapt new technologies like fire and marry outside the immediate family. Ernest's well-travelled Uncle Vanya – ha – mistrusts new ideas and embodies a smug Britishness that deems European Neanderthals 'too meta-physical', and dismisses the creatures inhabiting the rest of the world as unevolved. *What We Did to Father* is also a twentieth-century cautionary tale about the effects of technology. When Ernest's father invents the bow and arrow, his children sense that the social order is threatened, so they kill and eat him.[57]

What We Did to Father was followed in 1963 by the much gentler *Stig of the Dump*, Clive King's beloved children's book, which has been in print ever since. The protagonist is a lonely boy named Barney. He and his elder sister Lou are spending the summer with their grandparents in rural Kent. While exploring the Downs, Barney discovers and befriends a grunting Palaeolithic character living in a chalk pit and names him Stig. Adults dismiss Barney's stories about Stig, from whom he learns to hunt and to stand up to bullies and thieves, and of the power of the imagination and friendship. In the final scene, Stig's tribe raise standing stones for a solstice rite, after which Barney and Lou wake from a dream, questioning whether they really have witnessed prehistoric creatures. The scene tied Stig concretely to Britain's prehistoric past and to the comic cartoon tradition in which it had long been evoked. Stig's disappearance at the novel's end recalled Arthurian legend in the suggestion that he might still inhabit the Downs, giving readers the magical sense that friendly cave men will once again appear when needed.[58]

Lewis and King reasserted prehistory as a fruitful era from which to make intelligent satires of the contemporary world. The idea was cemented in the global imagination by *The Flintstones*, the first prime-time animated television programme and one of the most successful cartoons ever. The series, which was first aired in America from 1960 to 1966, was created by William Hanna and Joseph Barbera. They modelled the show on popular family situation comedies like *The Honeymooners*, but struggled to find an appropriate time period in which to set it until, as Barbera recalled:

> suddenly, it all started coming together. It was the objects, the gadgets, the everyday modern things translated into terms of the Stone Age that really drove the creation of the show. First we came up with the Stonewall piano, and that led to gag after gag. The cave was not the cold and brutal firelit hollow from which Victor Mature emerged in *One Million BC*, but a tastefully appointed suburban bungalow.[59]

Hanna had similarly ecstatic recollections about a drawing of a prehistoric gramophone – a bird that used its beak as a stylus – at which the animators 'all recognised the gag potential of a cartoon show that burlesqued modern conveniences ... adapting them to the Stone Age'. Such sight gags would be reinforced with sophisticated word play that appealed to adults.[60] The words could have been spoken by E.T. Reed seventy years earlier.

Advertisements announcing the first episode called *The Flintstones* 'a satire on modern suburban life', while the show's theme song proclaimed it told the story of a 'modern stone-age family'.[61] The latter almost perfectly distilled the cave man concept, by encapsulating the interplay between the prehistoric and contemporary worlds on which the comedy turns. The Flintstones are a modern, nuclear Western family inhabiting a prehistoric past that is really the affluent United States of the 1960s. The satire is broad. Fred Flintstone and his neighbour Barney Rubble hold lower-middle-class jobs in a quarry – where else would cave men work? Their homemaker wives Wilma and Betty are smarter but shrewish. The characters wear archaic versions of modern dress and live in Bedrock, a sunny suburb where life revolves around the drive-in, the service club and the bowling league. Archaic technology includes the family car that is propelled by the occupants' feet, a brontosaurus punch-clock and a mammoth that serves as the kitchen tap, while celebrity culture is sent up by the stars who appear under punning names like Gary Granite (Cary Grant) and Ann-Margarock (Ann-Margaret).[62]

The Flintstones was broadcast in dozens of countries along with a barrage of merchandising, like the children's vitamins manufactured by the pharmaceuticals company that partially financed the programme. The original episodes have been re-aired constantly for over fifty years, along with spin-offs and live action films, making Fred Flintstone the most recognisable cave man of all time.[63] The *Flintstones*' resemblance to Reed's vision of a prehistoric era inhabited by Western Europeans alone also reflected the virtual absence of positive minority characters on American television in the 1960s. But shortly after the programme was cancelled, Barbera proposed making a new one about a black family who lived in Bedrock named the Blackstones. Such a series might have commented on the contemporary Civil Rights and Black Power movements. Instead, it triggered racist unease similar to that with which Americans had greeted gorillas a century earlier. Much less significantly, the idea of a middle-class suburban black family cut across the grain of virtually every previous depiction of comic prehistory. Cave men are white. Television executives were completely uninterested, telling Barbera that such a series would be 'too provocative'; an unsubtle euphemism for their anxieties.[64]

Most subsequent evocations of prehistory have been compared with the *Flintstones*. The first of these was the deeply unfunny 1966 American television programme *It's About Time*, the Space Race story of two astronauts who travel faster than the speed of light and end up one million years in the past. What passes for comedy turns on the cave men's malapropisms and ungrammatical speech and the odd contemporary allusion, such as when the astronauts disguise themselves with hide clothing and unkempt wigs, and one calls the other 'Ringo'. As audience numbers plummeted, the astronauts and the cave family that had befriended them were transported to present-day Manhattan, on the premise, as the show's creator explained, that 'cavemen in modern times have an infinite variety of things to react to'.[65] Subsequent American television programmes, like *Land of the Lost*, in which a father and his two children are transported to a tropical planet inhabited by hominids, dinosaurs and aliens, or Hanna-Barbera's *Captain Caveman and the Teen Angels*, about a furry club-carrying blob, have all attempted to recapture *The Flintstones*' popularity.[66]

British filmmakers finally returned to the cave man genre in 1966 when Hammer Studios, the legendary home of low-budget horror films, invested £400,000 in a remake of *One Million BC* entitled *One Million Years BC*. Publicity photographs of the female lead, the American starlet Raquel Welch, clad in a leather bikini are among the iconic images of 1960s cinema. These demonstrate that the studio had forsaken the moral purpose of Hal Roach's original for B-movie sensibilities. The film opens with an ominous narration about a 'hard, unfriendly world' which is inhabited nonetheless by hyper-sexualised women with elaborate coiffures and tight-fitting clothing. The film's popularity in both Britain and the United States caused Hammer to make the eminently forgettable follow-up *Prehistoric Women*, and inspired *Voyage to the Planet of Prehistoric Women*, an incredibly cheap American production that spliced a redubbed Soviet space film with new scenes of pterodactyl-worshipping blonde, bikini-clad aliens led by Mamie Van Doren, a downmarket Marilyn Monroe whose career had long since peaked.[67]

Prehistoric films proliferated in Britain and the United States in the 1970s, beginning with *Trog*, in which the aged star Joan Crawford discovers an ancient hominid in the English countryside. In the laughably awful *Skullduggery* of that same year Burt Reynolds, whose career was on the opposite trajectory to Crawford's, finds a tribe of missing links enslaved in the wilds of Borneo and fights a court case to prove their humanity. Hammer also returned to the genre that year with *When Dinosaurs Ruled the Earth*, a titillation film dressed up with a weighty declaration about exploring the beginnings of love, hate and fear. It was followed in 1975 by *The Land that Time Forgot*, an expensive production in which First World War sailors venture to an island inhabited by dinosaurs and proto-humans. An erupting volcano strands the group's leader, fortunately for the studio, if not for the character, since it ushered in the sequel *The People that Time Forgot*, in which a rescue party is helped by an articulate cave woman with luxuriant hair and revealing leather clothes. That same year, the far less accomplished American film *Planet of Dinosaurs* married prehistory to a post-*Star Wars* fascination with science fiction. The crew of a spaceship land on a planet inhabited by dinosaurs, gradually trading their uniforms for loincloths and bikinis. Eventually it becomes clear that, as in *Teenage Caveman*, this is a post-apocalyptic earth, and the crew must procreate to begin a new cycle of evolution.[68]

Britain's final fascination with comic prehistory took place in the magazine *Punch*, in which the cave man had first appeared in the 1890s. In the 1970s, the magazine published over 200 cartoons tracing the adventures of the Neolithic hunter Stanley, who is described as a 'large nosed, short sighted neurotic little no-hoper'.[69] Stanley was drawn by the New Zealander Murray Ball, whose sense of humour was markedly different from Reed's. He satirised contemporary urban life and relationships. Ball could not have mimicked Reed too closely: the complacent sense of British greatness had not survived the world wars, the end of empire or 1960s iconoclasm.

Other projects tried to reimagine prehistory as an era from which serious lessons could be drawn. For instance, the director Stanley Kubrick's 1968 film *2001: A Space Odyssey* opens with one of the most famous sequences in cinema, entitled the 'Dawn of Man', in which grunting, simian proto-humans inhabit a sere landscape. When one of them wields an animal's femur as a weapon, they are transformed into violent and aggressive carnivores. The scene is scored by Richard Strauss's *Also sprach Zarathustra*, music inspired by Friedrich Nietzsche's novel of the same name, which posits that humans are an intermediary race between apes and the *Übermensch*, the so-called superman. The 1981 French-American film *Quest for Fire* used scientific evidence to accurately recreate prehistory, while *Iceman* of 1984 told the tale of a Cro-Magnon hunter who is discovered frozen in a glacier and revived and studied by scientists.[70]

A more heroic and less sexualised role for cave women was explored in literature through Jean M. Auel's 1980 novel *Clan of the Cave Bear*, which introduced Ayla, a Cro-Magnon shaman, healer, hunter and master of fire. She is also Eve, the progenitor of modern humans. Auel's success – *Clan* and its sequels have sold more than forty million copies – challenged masculine conceptions of prehistory. Her strong protagonist is more attuned to the aspirations of contemporary women than previous depictions of prehistoric females. The book was filmed in 1986 in an expensive production that strove for scientific accuracy, and downplayed Ayla's sexuality through a costume that was far less revealing than those worn by most recent cinematic cave women. *Clan* was a financial and critical disappointment.[71] Cave women still await their screen champion.

Such sober recreations of the ancient past are entertaining, but hard to love. Comedy continues to be a far surer tone for prehistory, from the 1981 film *Caveman*, in which a Cro-Magnon Ringo Starr courts an ultra-glamorous bikini-clad woman, to Mel Brooks's send-up of Kubrick in *History of the World Part 1* and the southern California high school angst of *Encino Man*, in which a pair of teenagers pass off a defrosted cave man as an exchange student from Estonia, a suitably punning country. Extremely cheap cave man films have appeared throughout. They generally combine action and sexual fantasy under unsubtle titles like *A Nymphoid Barbarian in Dinosaur Hell* (1990).[72] More ambitious productions have appeared of late, from gentle comedies like *Year One* to animated features like *Ice Age* and the live-action adventure of *10,000 BC*.[73] Television has had an equally ongoing fascination with comic cave men, from Phil Hartman's appearances as Keyrock, the verbose and materialistic 'Unfrozen caveman lawyer' on the programme *Saturday Night Live* in the 1990s, to the advertisements for the car insurance company Geico that have since 2004 featured well-spoken, middle-class Neanderthals.[74] The *Jurassic Park* films may not involve cave men, but their huge success signals that Hollywood will mine prehistory for many years to come.[75]

Almost a century separates E.T. Reed's cartoon of H.G. Wells and Ray Lankester and *The Croods*, the immensely successful 2013 animated feature which has already spawned cinematic and television sequels. Yet this American film about a cave family with overprotective parents and surly children shows how little popular cultural conceptions of the ancient past have changed in that time. The Croods inhabit a prehistoric world that is a backward projection of our own, peopled with Western humans whose thoughts, aspirations, emotions, family structures and interpersonal relations are modern.[76] Reed's prehistoric drawings were never particularly personal or intimate, and it is hard to imagine him creating cartoons that dissected family relationships. But Frederick Burr Opper first popularised cave men in the United States by reimagining them from this perspective. More recent evocations of prehistory have travelled easily between these poles of broad social commentary and intimate family life: shifting perspectives that audiences comprehend and navigate easily. Prehistory's appeal, to creators and audiences alike, lies primarily in its declaration that humans and Western society have been immutable through time. This does not challenge Anglo-Saxon, male audiences who see themselves and their social dominance projected on screen. Women, indigenous peoples and ethnic minorities have rarely if ever been accorded appropriate space in this imagined ancient past. Only the most intransigent opponents of science link *The Croods* and *The Flintstones* to palaeontological evidence and Charles Darwin's theories. Fred Flintstone looks, acts and sounds far too much like a modern American. His prehistoric world is intimately familiar, while also being sufficiently distant and different for audiences to laugh, at themselves, unselfconsciously.

Notes

1 The quotation is from H.G. Wells, *The Outline of History*, vol. I (London: Cassell & Co., 1920), p. 50. See also Matthew Skelton, 'The paratext of everything: constructing and marketing H.G. Wells's *The Outline of History*',

Book History, 4 (2001), pp. 237–48; and David C. Smith, *H.G. Wells: Desperately Mortal, a Biography* (New Haven: Yale University Press, 1986), pp. 249–58.
2 'Who says I'm not contemporary with man?', *Bystander* (24 December 1919), p. 1039; see also 'Wells goes back 800 million years to begin his story of life and man', *New York Tribune* (18 January 1920), p. 7.
3 Richard Overy, *The Morbid Age: Britain between the Wars* (London: Allen Lane, 2009), pp. 9–49.
4 'Personal ascendancy', *Burnley News* (1 November 1919), p. 3; 'Prehistoric Percy's present', *Burnley News* (28 November 1925), p. 12; 'Christmas productions', *Stage* (25 December 1919), p. 14; 'Bedford shopping carnival', *Bedfordshire Times* (8 October 1920), p. 7; 'Bedford School', *Bedfordshire Times* (8 July 1921), p. 8.
5 See for instance 'The playhouse', *Folkestone, Hythe, Sandgate and Cheriton Herald* (26 July 1919), p. 9; 'The "Prehistoric" as seen at the Albert Hall', *Illustrated London News* (8 January 1921), pp. 44–5; 'The pedlers', *Stage* (16 June 1921), p. 11; 'A dress pageant', *Western Morning News* (26 November 1921), p. 4; 'Fancy dress ball at Barnstaple', *North Devon Journal* (26 January 1922), p. 5; 'Biddenham', *Bedfordshire Times* (3 March 1922), p. 3; 'Brentwood fete', *Chelmsford Chronicle* (11 August 1922), p. 7; 'Prehistoric man at Hull carnival', *Hull Daily Mail* (27 July 1923), p. 3; 'Lowerhouse C.C. Pars', *Burnley Express* (2 February 1924), p. 11; 'Musings on motoring', *Taunton Courier* (12 August 1925), p. 5; 'Aston Clinton fete', *Bucks Herald* (15 August 1925), p. 10; 'Manchester Palace', *Stage* (28 May 1936), p. 2; 'Chelmsford carnival', *Chelmsford Chronicle* (9 July 1937), p. 2; 'University students in gala show', *Aberdeen Journal* (16 April 1938), p. 8.
6 'The railroad', *Falkirk Herald* (17 September 1930), p. 6; National Railway Museum, York, image no. 10449481.
7 'Wiveliscombe carnival', *Taunton Courier* (27 November 1929), p. 8. See also 'For charity', *Buckingham Advertiser and Free Press* (21 June 1930), p. 5.
8 'Venue problem for cavalcade of sport', *Gloucestershire Echo* (13 January 1937), p. 6; 'Scenes at Gloucester cavalcade of sport', *Gloucester Journal* (24 July 1937), p. 1.
9 Tammy M. Proctor, 'On my honour: Guides and Scouts in interwar Britain', *Transactions of the American Philosophical Society*, 92:2 (2002), p. 17.
10 Ibid., pp. 13–19 and 89–93.
11 'Scouts' jamboree', *Western Daily Press* (31 July 1920), p. 7.
12 'Work hard and play hard', *Aberdeen Daily Journal* (4 October 1920), p. 5.
13 See for instance 'Prehistoric Aberdonians', *Aberdeen Daily Journal* (4 May 1921), p. 3; 'Wishaw news', *Motherwell Times* (14 June 1929), p. 8; 'Wishaw news', *Motherwell Times* (28 June 1929), p. 8; 'Derby Scout notes', *Derby Telegraph* (1 July 1930), p. 6; 'Derby's Mayor a good Scout', *Derby Telegraph* (14 July 1930), p. 4; 'Prehistoric Scouts', *Dundee Courier* (1 August 1930), p. 5; 'Caterham carnival', *Surrey Mirror* (29 August 1930), p. 8; 'Grand howl by

Wolf Cubs', *Cheltenham Chronicle* (30 May 1931), p. 6; and 'Fetes and parties everywhere', *Derby Telegraph* (11 July 1931), p. 1.

14 Australia: 'A Scouts' muster', *Sydney Morning Herald* (21 April 1924), p. 8; 'Boy Scouts', *South Coast Times and Wollongong Argus* (9 December 1932), p. 22; 'Scouts and Scouting', *World's News* (Sydney) (22 March 1933), p. 26; 'Always a Scout', *Argus* (Melbourne) (7 January 1935), p. 8; 'Scouts' bonfire and torchlight parade', *Manning River Times* (New South Wales) (3 June 1953), p. 3. United States: 'Chevy Chase', *Evening Star* (Washington, DC) (28 May 1922), p. 6. New Zealand: 'With the Scouts', *Auckland Star* (18 September 1913), p. 8; 'Scout notes', *Evening Post* (Wellington) (26 November 1924), p. 15.

15 See for instance '12th March sports', *Gloucester Advocate* (Australia) (5 March 1921), p. 2; 'Gallery art students revel', *Table Talk* (Melbourne) (13 July 1922), p. 36; Syd Nicholls, 'Prehistoric footballer', *Sydney Evening News* (25 June 1923), p. 4; 'Gala night at maison de luxe', *Table Talk* (Melbourne) (26 July 1923), p. 12; 'Art students' annual ball', *Table Talk* (Melbourne) (3 July 1924), p. 16; and 'The Christmas *Strand*', *Telegraph* (Brisbane) (27 December 1934), p. 4. For advertisements see 'Inaction no longer menaces health', *Western Mail* (Perth) (15 November 1918), p. 42; 'Suppose you were a cave man', *Brisbane Courier* (30 December 1918), p. 8; 'Plimsoll's', *Press* (Canterbury, New Zealand) (21 October 1921), p. 12; 'Laurel kerosene', *Observer* (Adelaide) (10 May 1924), p. 61; 'Laurel kerosene', *Northern Advocate* (New Zealand) (26 August 1924), p. 3; 'Andrew's liver salt', *Press* (Canterbury, New Zealand) (9 January 1926), p. 6; 'A pebble was the cave man's candy', *Daily Telegraph* (Tasmania) (27 August 1927), p. 3; and 'A pebble was the cave man's candy', *Auckland Star* (10 September 1927), p. 34.

16 'Cave life or civilization?', *Boy's Life – the Boy Scouts Magazine* (April 1916), p. 19; 'Prehistoric peeps', *Life* (13 November 1919), p. 827; 'How about your roof?' *Rockingham Post-Despatch* (North Carolina) (21 October 1920), p. 10.

17 William J. Fielding, *The Caveman Within Us* (New York: Dutton, 1922).

18 'Husbands organize to protect rights', *New York Times* (9 November 1922), p. 14; 'Novel Goldwyn exploitation gets nation-wide attention', *Exhibitors Herald* (25 November 1922), p. 49.

19 'Cave-man stuff', *Variety* (2 July 1924), p. 5.

20 See for instance 'Cave boy stuff goes very bad', *Chicago Tribune* (22 October 1916), p. 4; 'No "rough stuff", cave man if you marry a Girl Scout', *Richmond Times-Dispatch* (Virginia) (4 March 1920), p. 1; and 'She first jailed then married her fierce cave-man lover', *Tulsa Daily World* (17 September 1922), unpaginated. See also Modern cave man', *Western Gazette* (Australia) (21 December 1923), p. 9; 'Cave man conduct', *New Zealand Truth* (5 April 1924), p. 12; and 'Cave man', *Truth* (Queensland) (7 July 1929), p. 19.

21 Edward B. Davis, 'Fundamentalist cartoons, modernist pamphlets, and the religious image of science in the Scopes era', in Charles L. Cohen and Paul S. Boyer (eds.), *Religion and the Culture of Print in Modern America* (Madison: University of Wisconsin Press, 2008), p. 191.

22 Jeffrey P. Moran, *American Genesis: The Antievolution Controversies from Scopes to Creation Science* (New York: Oxford University Press, 2012), pp. 16–20; Davis, 'Fundamentalist cartoons', pp. 175–82.
23 The quotation is from Constance Areson Clark, 'Evolution for John Doe: pictures, the public, and the Scopes trial debate', *Journal of American History*, 87:4 (2001), p. 1303. See also Judith C. Berman, '"Bad hair days in the Paleolithic": modern (re)constructions of the cave man', *American Anthropologist*, 101:2 (1999), p. 300; and Constance Areson Clark, *God – or Gorilla: Images of Evolution in the Jazz Age* (Baltimore: Johns Hopkins University Press, 2008), p. 4.
24 *The First Circus* (1921), Library of Congress, 00694026.
25 Moran, *American Genesis*, pp. 19–23; Julie Homchick, 'Objects and objectivity: the evolution controversy at the American Museum of Natural History, 1915–1928', *Science & Education*, 19:4–5 (2010), pp. 490 and 500–1; Ronald L. Numbers, *The Creationists: From Scientific Creationism to Intelligent Design* (Cambridge, Massachusetts: Harvard University Press, 2006). For the film see 'Evolution', *Exhibitor's Trade Review* (8 August 1925), p. 39.
26 G.K. Chesterton, 'The everlasting man', in *The Collected Works of G.K. Chesterton*, vol. II (San Francisco: Ignatius Press, 1986), p. 159.
27 The quotation is from Maria Wyke, 'Silent laughter and the counter-historical: Buster Keaton's "Three Ages"', in Pantelis Michelakis and Maria Wyke (eds.), *The Ancient World in Silent Cinema* (Cambridge: Cambridge University Press, 2013), p. 286. See also Daniel Moews, *Keaton: The Silent Features Close Up* (Berkeley: University of California Press, 1977), pp. vii and 1–10; and Marion Meade, *Buster Keaton; Cut to the Chase* (New York: Harper Collins, 1995), pp. 133–4.
28 Buster Keaton (dir.), *Three Ages* (1923); Meade, *Buster Keaton*, p. 134; Moews, *Keaton*, pp. 323–4; and Wyke, 'Silent laughter', pp. 292–3.
29 Wyke, 'Silent laughter', pp. 277–84.
30 Meade, *Buster Keaton*, p. 136; 'The Majestic', *Yorkshire Evening Post* (25 June 1923), p. 1; 'The Caley', *Edinburgh Evening News* (25 June 1923), p. 1; 'Metro film showing attended by royalty', *Motion Picture World* (14 July 1923), p. 162.
31 'Film gossip', *Nottingham Evening Post* (18 January 1919), p. 4.
32 See for instance 'Stoll to build another', *Variety* (18 July 1919), p. 4; George Robey, 'My animated rest cure', *Pictures and Picturegoer* (December 1923), p. 64.
33 Bryony Dixon, 'The ancient world on silent film: the view from the archive', in Michelakis and Wyke (eds.), *Ancient World*, pp. 30–1.
34 E.T. Reed, 'Prehistoric peeps', *Punch* (20 October 1894), p. 189.
35 A.E. Coleby (dir.), *The Prehistoric Man* (1924), British Film Institute.
36 Randy Skretvedt and Jordan R. Young, *Laurel and Hardy: The Magic behind the Movies* (Beverly Hills: Moonstone Press, 1987), pp. 89–90.
37 Hal Roach (dir.), *Flying Elephants* (1928).
38 See for instance Leonard Dove, 'Cripes, it's the wife!', *New Yorker* (1 December

1934); and Rea Irvin, 'Won't you step in and look at my etchings?', *New Yorker* (29 May 1937).
39 Nicholas Ruddick, *The Fire in the Stone: Prehistoric Fiction from Charles Darwin to Jean M. Auel* (Middletown: Wesleyan University Press, 2009), p. 78.
40 Chuck Jones (dir.), *Daffy Duck and the Dinosaur* (1939).
41 Richard Lewis Ward, *A History of the Hal Roach Studios* (Carbondale: Southern Illinois University Press, 2005), pp. 109–11; Hal Roach (dir.), *One Million BC* (1940).
42 Robert Clampett (dir.), *Prehistoric Porky* (1940); Jack Kinney (dir.), *The Art of Self Defence* (1941).
43 William Beaudine (dir.), *The Ape Man* (1943); Phil Rosen (dir.), *Return of the Ape Man* (1944).
44 Seymour Kneitel (dir.), *Prehysterical Man* (1948).
45 Jules White (dir.), *I'm a Monkey's Uncle* (1948); Jules White (dir.), *Stone Age Romeos* (1955).
46 Robert McKimson (dir.), *Prehysterical Hare* (1958).
47 Sam Arkoff with Richard Trubo, *Flying through Hollywood by the Seat of my Pants* (New York: Birch Lane Press, 1992), p. 58.
48 Ibid., pp. 41–2; Blair Davis, *The Battle for the Bs: 1950s Hollywood and the Rebirth of Low-Budget Cinema* (New Brunswick, New Jersey: Rutgers University Press, 2012), pp. 22–5, 36–42 and 78.
49 Katia Busch, 'Nus de style préhistorique', in *Vénus et Caïn: Figures de la préhistoire, 1830–1930* (Bordeaux: Éditions de la réunion des musées nationaux, 2003), pp. 144–51; 'New Orleans', *Variety* (19 March 1915), p. 29.
50 Gregg C. Tallas (dir.), *Prehistoric Women* (1950).
51 Arkoff, *Flying*, pp. 72–3; Roger Corman with Jim Jerome, *How I Made a Hundred Movies in Hollywood and Never Lost a Dime* (New York: Random House, 1990), pp. 56–7; Irving Rubine, 'Boys meet ghouls, make money', *New York Times* (16 March 1958), p. X7.
52 Roger Corman (dir.), *Teenage Caveman* (1958).
53 Arch Hall Sr (dir.), *Eegah* (1962).
54 Osbert Lancaster, 'Draynefleet revealed', in *The Littlehampton Saga* (London: Methuen, 1984), p. 87.
55 Parliamentary Debates (Hansard), vol. 466 (London: His Majesty's Stationery Office, 1949), 5 July 1949, col. 1955.
56 Clive Gamble, 'Foreword', in Stephanie Moser, *Ancestral Images: The Iconography of Human Origins* (Ithaca: Cornell University Press, 1998), p. ix.
57 Roy Lewis, *The Evolution Man: or How I Ate my Father* (New York: Pantheon Books, 1960), passim.
58 Clive King, *Stig of the Dump* (London: Puffin, 2003), passim. See also 'Choosing a modern classic', *Times Literary Supplement* (25 November 1983), p. 1311.
59 The quotation is from Joseph Barbera, *My Life in 'Toons: From Flatbush to Bedrock in Under a Century* (Atlanta: Turner Publishing, 1994), p. 5. See also

Bill Hanna with Tom Ito, *A Cast of Friends* (Dallas: Taylor Publishing, 1996), pp. 110–12.
60 Hanna, *Cast*, pp. 112–14. The quotation is from p. 112.
61 Fred Danzig, '"Flintstones" TV hoax of season', *Desert Sun* (California) (7 February 1961), unpaginated.
62 Jeff Lenburg, *William Hanna and Joseph Barbera: The Sultans of Saturday Morning* (New York: Chelsea House, 2011), p. 98.
63 Ibid., p. 106; Mike Barnes, 'Toon legend Barbera dies', *Hollywood Reporter – International Edition* (19 December 2006), pp. 4 and 57; Rich Cohen, 'Darkness at the edge of Bedrock', *Rolling Stone* (30 June 1994), p. 63; Lacey Rose and Matthew Belloni, 'What killed Seth Macfarlane's *Flintstones* remake?', *Hollywood Reporter* (4 May 2012), p. 8.
64 Joseph Barbera, as quoted in Lenburg, *William Hanna*, p. 113.
65 Sherwood Schwartz (creator), *It's About Time* (1966–67).
66 Allan Foshko, Marty Krofft and Sid Krofft (creators), *Land of the Lost* (1974); Joe Ruby and Ken Spears (creators), *Captain Caveman and the Teen Angels* (1977).
67 Don Chaffey (dir.), *One Million Years BC* (1966); Michael Carreras (dir.), *Prehistoric Women* (1967); Derek Thomas (dir.), *Voyage to the Planet of Prehistoric Women* (1968).
68 Freddie Francis (dir.), *Trog* (1970); Gordon Douglas and Richard Wilson (dirs.), *Skullduggery* (1970); Val Guest (dir.), *When Dinosaurs Ruled the Earth* (1970); Kevin Connor (dir.), *The Land that Time Forgot* (1975); Kevin Connor (dir.), *The People that Time Forgot* (1977); James K. Shea (dir.), *Planet of Dinosaurs* (1977).
69 Murray Ball, *Stanley* (Lower Hutt: INL Print, undated), p. 8.
70 Jean-Jacques Annaud (dir.), *Quest for Fire* (1981); Fred Schepisi (dir.), *Iceman* (1984).
71 Michael Chapman (dir.), *The Clan of the Cave Bear* (1986); Ruddick, *Fire*, pp. 84–5.
72 Carl Gottlieb (dir.), *Caveman* (1981); Mel Brooks (dir.), *History of the World Part 1* (1981); Brett Piper (dir.), *A Nymphoid Barbarian in Dinosaur Hell* (1990); Les Mayfield (dir.), *Encino Man* (1992); Beverley Gray, *Roger Corman: An Unauthorized Biography of the Godfather of Indie Filmmaking* (Los Angeles: Renaissance Books, 2000), pp. 192–3.
73 Chris Wedge and Carlos Saldanha (dirs.), *Ice Age* (2002); Harold Ramis (dir.), *Year One* (2009); Roland Emmerich (dir.), *10,000 BC* (2008).
74 Phil Hartman, 'Unfrozen caveman lawyer', *Saturday Night Live*, various episodes (1991–96); Joe Lawson (creator), Geico cavemen (2004–present).
75 Steven Spielberg (dir.), *Jurassic World* (2015).
76 Kirk De Micco and Chris Sanders (dirs.), *The Croods* (2013).

Conclusion

In July 2015, the *Times Literary Supplement* published a featured review of the book *Political Descent: Malthus, Mutualism and the Politics of Evolution in Victorian England*, which, as its punning title suggests, explores how Charles Darwin's theories were used to justify nineteenth-century radicalism.[1] It is a weighty academic enquiry from a leading university press. Yet when the *Supplement*'s editors selected an image for the issue's cover, they overlooked myriad photographs of the grave and gloomy Darwin or over-populated Victorian slums. Instead, they chose E.T. Reed's 1894 *Punch* cartoon 'Opening of the primeval Royal Academy', which depicts the artistic lions of late Victorian Britain vying, often violently, to hang their rock drawings most advantageously on the walls of Burlington House.[2] The cartoon is not well known. It has no overt political message. And it is not about evolution. Reed projected the Royal Academy Summer Exhibition, Britain's most socially prestigious art show, into prehistory.

The cartoon is also apposite because it does not immediately strike the *Supplement*'s readers as a Victorian period-piece, unlike, say, John Tenniel's more famous illustrations for *Alice in Wonderland*. Modern viewers instinctively recognise Reed's conceit that the prehistoric world was an archaic version of our own, thanks to familiarity with more recent cartoons like *The Flintstones* that adapted his comic sensibilities. Reed projected late Victorian and Edwardian British institutions, ideals and events into prehistory in order to satirise them. He knew that such a fanciful context would make the scenes both familiar and distant, giving him wide a scope of subjects to tackle and allowing his audiences to laugh at themselves unselfconsciously. Generations of illustrators, writers and performers have subsequently used cave men and their ancient world to send up contemporary society.

As the decidedly upper-middle-class setting of the Royal Academy demonstrates, such satires are best wielded by those like Reed, whose membership in social, cultural and political elites gives them the confidence to mock their society's values, customs and institutions. Reed was no iconoclast. He sought only to gently prod imperial Britain. It is also important that he did so in the pages of *Punch* magazine, which epitomised an assured, self-deprecating British humour that appealed equally at home and throughout the empire. *Punch* and its many imitators had viewed the furore about evolution as a great joke at the expense of those whose scientific or religious beliefs were too passionate and grave. Prehistoric comedy was less biting, in part because it upheld racial, national and imperial prejudices.

Comic evocations of prehistory have been immensely popular for over a century, but it is important to calibrate and contextualise them. Showmen had long used vague pseudo-evolutionary, racist ideas to create exciting stories for exhibits of so-called freaks and indigenous peoples, reflecting a generalised public understanding of evolution, apes and monkeys. In the mid-nineteenth century evolution became a touchstone for religious, national and racial insecurities in part because Darwin, its greatest exponent, was not a political radical, hack journalist or humbugging showman. He was an extremely talented scientist, from an eminent family whose ideas could not be easily dismissed. Reactions to Darwinian theories were amplified when linked to the comprehensive sexual and physical threat that gorillas seemed to represent. Could humans actually be descended from such a dark African beast? Apostles preached the contradictory claims of science and faith. Humour was a more widespread but more oblique response. *Punch*'s 1861 'Mr Gorilla' cartoon presented a farcical, nightmare vision of the world turned upside down, the temporary social inversion associated with pre-modern carnivals. But unlike those seasonal fairs, the cartoon jokily forewarned of a permanent realignment based on the physiological proximity of humans and apes. Laughing at Mr Gorilla required a generalised knowledge of evolutionary ideas, but not wholesale belief in or rejection of them. Myriad songs, plays, pantomimes, colloquial expressions and cartoons subsequently charted the contested territory between apes and humans. The idea that a gorilla in white tie and tails might actually be passed off as an English gentleman was both deeply threatening and absurd, since costumed acrobats suggested that the creature was only a dextrous, dumb animal.

Evolution's implicit threat ultimately made it a limiting context in which to imagine ancient humans. Gorillas and missing links could never be entirely divorced from earlier traditions of evolutionary showmanship. They were fundamentally monsters, no matter how completely they were dressed up in satire. Prehistory provided a far more expansive artistic and imaginative space. The modern humans who populated this era reflected popular scientific knowledge about the first Europeans, without the degrading proximity to apes. Cave men posed no racial, physical or existential threats to modern society. Their fanciful world never existed, though it is extremely comforting to think that it did, because it suggested that modern humans had ever been thus. The conceit is not predicated on believing that our simian ancestors had walked out of the African savannah aeons ago. Instead, it posits that fair-skinned Anglo-Saxons had coexisted in some unexplained way with dinosaurs, extinct mammals and archaic versions of contemporary institutions, events and technology.

Lest we condescend too severely at this Victorian vision of an immutable order, we should remember that the changing seasons are celebrated at Stonehenge each year by crowds of self-styled druids. English Heritage, which manages the monument for the nation, allows these pagans access to the megaliths to perform a rite that was invented in the early twentieth century. Like the world of Reed's prehistoric peeps, it is a fanciful modern evocation of a little-understood epoch in Britain's ancient past. Yet in permitting such ceremonies, however begrudgingly, the government implicitly legitimates this imagined view of history. How different is this from Victorian and Edwardian historical parades and pageants that included cave men?

In a study of cinematic depictions of the ancient world, Jeffrey Richards argued that 'historical films are always about the time in which they are made and never about the time in which they are set'.[3] So too with prehistory. In the 1890s, Reed drew prehistoric cartoons for the upper middle class to which he belonged. Male members of this elite saw themselves as English gentlemen, standing at the apex of a finely graded hierarchy of class, culture and race within Britain, the empire and the globe. Reed's humour suggested that this stratification had been eternal, in part by virtually erasing other races, cultures and ethnicities from the ancient past. This conception of a prehistory based on social institutions rather than biology also made the era palatable in the United States. But class and race can never be entirely separated, given the intimate correlation between skin colour, wealth and social status. This is replicated in prehistory, where heroic and technologically advanced cave men are almost always depicted as light-skinned, slim and fair-haired, while their darker, dumber, heavier and more hirsute opponents appear to be far closer to gorillas. Women are equally absent or degraded in this deeply conservative view of the ancient past.

The reaction to Reed's first peeps cartoons showed that they were unprecedented and extraordinary. They were almost immediately incorporated into and transmitted throughout a mass culture that was international, imitative and ever searching for new sensations. This culture virtually obliterated divisions between the ideas that were presented on stage, page and screen. Frederick Burr Opper, Lawson Wood, George Robey, Buster Keaton and many less talented people saw the potential financial rewards of adapting Reed's vision of prehistory. Local, national and international businesses also perceived that they could sell goods by reflecting such passions back to consumers. The commercialisation of cave men showed that they had journeyed from the exotic to the everyday. Film and television are self-evidently important means through which cultural characters travelled the globe, though searchable databases of newspapers and magazines reveal analogous, if less obvious, earlier routes of cultural transmission. Broadcast technologies – from printed words transported by train and ship to wireless, film, television and the Internet – have become faster over the past century, but their reach is not much greater than that of the devices that existed in 1900. Cave men had been established in global popular culture by 1914, thereby providing soldiers from throughout the English-speaking world with a casual means of communicating with one another.

Cave men have limitations. The prehistoric peeps cartoons were a minor part of E.T. Reed's overall output. The Hanna-Barbera studio had many successes apart from *The Flintstones*. Writers as eminent as Rudyard Kipling and H.G. Wells explored prehistory's fictional possibilities but moved on. For performers like Robey and Keaton, the cave man was one in a longer roster of characters they played. Those who had nothing but cave men to offer were soon forgotten or relegated, as acrobatic gorillas had once been, to the poorer margins of the entertainment industry.

Constant exposure to cave men in mass culture has also created specific expectations in contemporary audiences. Cave men are comic. They are Anglo-Saxon. They are middle-class. They use archaic technology. They battle dinosaurs and extinct mammals. They court with their clubs. Their women are young, buxom and sexually voracious. Scientifically accurate depictions of prehistory in museums, documentaries and educational tools are entertaining, but hard to love. Serious attempts to recreate the prehistoric past also tend to shunt cave men aside in favour of dinosaurs, and especially the aggressive carnivorous monster *Tyrannosaurus rex*. Lugubrious herbivores like the brontosaurus – once dismissed but now rehabilitated – are nowhere near as popular. Generally, audiences do not expect dinosaurs to be funny. The 2015 instalment of the *Jurassic Park* film franchise, *Jurassic World*, demonstrated this, as the heroes battled a creature that had been specifically bred as a hybrid of aggressive traits in order to attract tourists.[4]

This conclusion opened by pointing to a modern reuse of Reed's pioneering images and now closes with a contemporary one that is implicitly indebted to him. In October 2015 the *New Yorker* published a drawing of two spear-carrying cave men clad in animal skins and gazing over a grassy vale. One asks the other, 'If you tasted like umami, where would you be hiding?', a question that satirises the effects of flavour-enhancers on the modern Western palate.[5] Black and white *New Yorker* cartoons, which incorporate verbal and visual puns, are the best modern equivalent of those that appeared in *Punch* over a century ago. The magazine is also published in the global cultural metropolis and distributed throughout the world. Readers far from Manhattan comprehend its contents because of a shared culture and idealised notions about New York. A nineteenth-century Torontonian, Sidneysider or Chicagoan would have had an equally good sense about life in London. This modern cartoon is therefore something of a prehistoric relic. Its comic sensibilities are rooted in mid-Victorian battles over evolution and in Reed's discovery that prehistory was a fertile era for satire. The image explicitly suggests that modern, middle-class, food-obsessed New Yorkers have always existed and, equally comforting, that cave man humour is eternal.

Notes

1 Gregory Raddick, 'Dismal destinies', *Times Literary Supplement* (3 July 2015), pp. 3–4. The book under review was Piers J. Hale, *Political Descent: Malthus, Mutualism and the Politics of Evolution in Victorian England* (Chicago: University of Chicago Press, 2015).
2 *Times Literary Supplement* (3 July 2015), cover. The cartoon is E.T. Reed, 'Prehistoric peeps', *Punch* (12 May 1894), p. 226.
3 Jeffrey Richards, *Hollywood's Ancient Worlds* (London: Continuum, 2008), p. 1.
4 Steven Spielberg (dir.), *Jurassic World* (2015).
5 Drew Dernavich, 'If you tasted like umami, where would you be hiding?', *New Yorker* (19 October 2015), p. 42.

Bibliography

Principal online databases of historic newspapers and magazines

The following stands in for a traditional exhaustive list of newspapers, magazines and printed ephemera consulted while researching this book. These databases were searched repeatedly using many different keywords over about three years. Most searches were done against the entire contents of a database. Throughout that time, new material was constantly added to databases, as it continues to be. So the list of titles searched was constrained only by the contents of a database on a given day.

British Newspaper Archive, www.britishnewspaperarchive.co.uk (last accessed 30 June 2016).
Canadiana.org, http://www.canadiana.ca/ (last accessed 30 June 2016).
'Chronicling America', Library of Congress, http://chroniclingamerica.loc.gov/ (last accessed 30 June 2016).
The Internet Archive, https://archive.org/index.php (last accessed 30 June 2016).
'Making of America', Cornell University and University of Michigan, http://ebooks.library.cornell.edu/m/moa/ (last accessed 30 June 2016).
Media History Digital Library, http://mediahistoryproject.org/ (last accessed 30 June 2016).
'Papers Past', National Library of New Zealand, https://paperspast.natlib.govt.nz/ (last accessed 30 June 2016).
'Trove', National Library of Australia, http://trove.nla.gov.au (last accessed 30 June 2016).
Welsh Newspapers Online, National Library of Wales, http://newspapers.library.wales/ (last accessed 30 June 2016).

Archival collections

British Library, London

Evanion Collection
Music Collections
Punch archives

Library of Congress, Washington, DC

Caroline and Erwin Swann collection of caricature & cartoon
Music copyright deposits collection
Prints and photographs division

University of Sheffield

Fairground Archives

Victoria and Albert Museum, London

Prints, drawings and paintings collection
Theatre and performance collection

Published sources

Adams, William Davenport, *A Book of Burlesque* (London: Henry, 1891).
Addison, Henry R., *The Blue-Faced Baboon* (London: John Dicks, 1884?).
Aguirre, Robert D., 'Exhibiting degeneracy: the Aztec children and the ruins of race', *Victorian Studies Association of Western Canada*, 29:2 (2003), pp. 40–63.
Allardt, Erik, 'Edward Westermarck: a sociologist relating nature and culture', *Acta Sociologica*, 43:4 (2000), pp. 299–306.
Altick, Richard, *The Shows of London* (Cambridge, Massachusetts: Harvard University Press, 1978).
Anderson, Patricia, *The Printed Image and the Transformation of Popular Culture* (Oxford: Clarendon Press, 1991).
Arkoff, Sam, with Richard Trubo, *Flying through Hollywood by the Seat of my Pants* (New York: Birch Lane Press, 1992).
Bailey, Peter, 'Conspiracies of meaning: music-hall and the knowingness of popular culture', *Past & Present*, 144 (1994), pp. 138–70.
—— *Leisure and Class in Victorian England: Rational Recreation and the Contest for Control, 1830–1885* (Toronto: University of Toronto Press, 1978).
—— (ed.) *Music Hall: The Business of Pleasure* (Milton Keynes: Open University Press, 1986).
Ball, Murray, *Stanley* (Lower Hutt: INL Print, undated).

Ballantyne, R.M., *The Gorilla Hunters* (London: T. Nelson and Sons, 1861).
Bannister, Robert C., *Social Darwinism: Science and Myth in Anglo-American Social Thought* (Philadelphia: Temple University Press, 1979).
Barbera, Joseph, *My Life in 'Toons: From Flatbush to Bedrock in Under a Century* (Atlanta: Turner Publishing, 1994).
Baring, Maurice, *With the Russians in Manchuria* (London: Methuen, 1905).
Barnum, Phineas Taylor, *The Autobiography of P.T. Barnum* (London: Ward and Lock, 1855).
Barrett-Hamilton, G.E.H., and H.O. Jones, 'A visit to Karaginski Island, Kamchatka', *Geographical Journal*, 12:3 (1898), pp. 280–99.
Beer, Gillian, *Darwin's Plots: Evolutionary Narrative in Darwin, George Eliot and Nineteenth Century Fiction* (Cambridge: Cambridge University Press, 1983).
Berman, Judith C., 'Bad hair days in the Paleolithic: modern (re)constructions of the cave man', *American Anthropologist*, 101:2 (1999), pp. 288–304.
Betts, John R., 'P.T. Barnum and the popularisation of natural history', *Journal of the History of Ideas*, 20:3 (1959), pp. 353–68.
Blake, Robert, *Disraeli* (London: Prion, 1998).
Blancke, Stefaan, 'Lord Monboddo's ourang-outang and the origin and progress of language', in Marco Pina and Nathalie Gontier (eds.), *The Evolution of Social Communication in Primates* (New York: Springer, 2014), pp. 31–44.
Boitard, Pierre, *Paris avant les hommes* (Paris: Passard, 1861).
Boyer, Deborah Deacon, 'Picturing the other: images of Burmans in imperial Britain', *Victorian Periodicals Review*, 35:3 (2002), pp. 214–26.
Bratton, J.S., 'English Ethiopians: British audiences and black-face acts, 1835–1865', *Yearbook of English Studies*, 11 (1981), pp. 127–42.
—— (ed.) *Music Hall: Performance and Style* (Milton Keynes: Open University Press, 1986).
Broks, Peter, 'Science, media and culture: British magazines, 1890–1914', *Public Understanding of Science*, 2:2 (1993), pp. 123–39.
—— 'Science, the press and empire: Pearson's publications, 1890–1914', in John M. Mackenzie (ed.), *Imperialism and the Natural World* (Manchester: Manchester University Press, 1990).
Brown, Jane E., and Richard Samuel West, 'William Newman (1817–1870): A Victorian cartoonist in London and New York', *American Periodicals*, 17:2 (2007), pp. 143–83.
Brown, Lucy, 'The treatment of the news in mid-Victorian newspapers', *Transactions of the Royal Historical Society*, 27 (1977), pp. 23–39.
Browne, Janet, 'Charles Darwin as a celebrity', *Science in Context*, 16:1 (2003), pp. 175–94.
—— 'Darwin in caricature: a study in the popularisation and dissemination of evolution', *Proceedings of the American Philosophical Society*, 145:4 (2001), pp. 496–509.
—— *Darwin's 'Origin of Species': A Biography* (Toronto: Douglas & McIntyre, 2006).
—— 'Squibs and snobs: science in humorous British undergraduate magazines around 1830', *History of Science*, 30 (1992), pp. 165–97.

Burkhardt, Frederick (ed.), *The Correspondence of Charles Darwin*, vol. XI (Cambridge: Cambridge University Press, 1999).
Burroughs, Edgar Rice, *The Eternal Lover* (Chicago: McLurg, 1925).
Busch, Katia, 'Nus de style préhistorique', in *Vénus et Caïn: Figures de la préhistoire, 1830–1930* (Bordeaux: Éditions de la réunion des musées nationaux, 2003), pp. 144–51.
Byron, Henry J., *Miss Eily O'Connor, A New and Original Burlesque Founded on the Great Sensation Drama of The Colleen Bawn* (London: Thomas Hailes Lacy, 1861).
Cannadine, David, *Ornamentalism: How the British Saw their Empire* (Oxford: Oxford University Press, 2001).
Cantor, Geoffrey, et al., *Science in the Nineteenth Century Periodical: Reading the Magazine of Nature* (Cambridge: Cambridge University Press, 2004).
'Cartoon', in *The Oxford English Dictionary*, vol. II (Oxford: Clarendon Press, 1978), p. 140.
Caudill, Edward, 'The bishop-eaters: the publicity campaign for Darwin and *On the Origin of Species*', *Journal of the History of Ideas*, 55:3 (1994), pp. 441–60.
—— 'The press and tails of Darwin: Victorian satire of evolution', *Journalism History*, 20:3 (1994), pp. 107–15.
Chaplin, Charles, *My Autobiography* (London: Bodley Head, 1964).
Chapman, James, *British Comics: A Cultural History* (London: Reaktion Books, 2011).
Chesterton, G.K., *The Collected Works of G.K. Chesterton*, vol. II (San Francisco: Ignatius Press, 1986).
Clark, Constance Areson, 'Evolution for John Doe: pictures, the public, and the Scopes trial debate', *Journal of American History*, 87:4 (2001), pp. 1275–1303.
—— *God – or Gorilla: Images of Evolution in the Jazz Age* (Baltimore: Johns Hopkins University Press, 2008).
—— '"You are here": missing links, chains of being and the language of cartoons', *Isis*, 100:3 (2009), pp. 571–89.
Clarke, Peter, *Hope and Glory: Britain 1900–1990* (London: Allen Lane, 1996).
Colby, Charles W., *Canadian Types of the Old Regime* (New York: Henry Holt, 1908).
Collins, Paul, 'Gorillas, I presume', *New Scientist Magazine*, 188:2519 (2005), pp. 46–7.
Collins, Philip, 'Some unpublished comic duologues of Dickens', *Nineteenth-Century Fiction*, 31:4 (1977), pp. 440–9.
Collins, Wilkie, *The Moonstone* (London: Tinsley Brothers, 1868).
Compton, Roy, 'A chat with Mr E.T. Reed', *Idler* (May 1896), pp. 505–6.
Conlin, Jonathan, *Evolution and the Victorians* (London: Bloomsbury, 2014).
Cook, James, *The Arts of Deception: Playing with Fraud in the Age of Barnum* (Cambridge, Massachusetts: Harvard University Press, 2001).
Corbin, John, *The Cave Man* (New York: Appleton, 1907).
Corman, Roger, with Jim Jerome, *How I Made a Hundred Movies in Hollywood and Never Lost a Dime* (New York: Random House, 1990).
Cotes, Peter, *George Robey: The Darling of the Halls* (London: Cassell, 1972).
Culbertson, Tom, 'The golden age of American political cartoons', *Journal of the Gilded Age and Progressive Era*, 7:3 (2008), pp. 276–95.
Cunliffe, Barry, 'Introduction: the public face of the past', in John D. Evans, Barry

Cunliffe and Colin Renfrew (eds.), *Antiquity and Man: Essays in Honour of Glyn Daniel* (London: Thames and Hudson, 1981), pp. 192–4.

Curtis, L.P., *Apes and Angels: The Irishman in Victorian Caricature* (Washington, DC: Smithsonian Press, 1997).

Daniel, Glyn, and Colin Renfrew, *The Idea of Prehistory* (Edinburgh: Edinburgh University Press, 1988).

Darwin, Charles, *On the Origin of Species* (London: John Murray, 1859).

Davidoff, Leonore, and Catherine Hall, *Family Fortunes: Men and Women of the English Middle Class, 1780–1850* (London: Hutchinson, 1987).

Davidson, Jane P., 'Henry A. Ward, "catalogue of casts of fossils" (1866) and the artistic influence of Benjamin Waterhouse Hawkins on Ward', *Transactions of the Kansas Academy of Science*, 108:3–4 (2005), pp. 138–48.

Davies, Christine, 'The English mother-in-law joke and its missing relatives', *Israeli Journal of Humour Research*, 1:2 (2012), pp. 12–39.

Davis, Blair, *The Battle for the Bs: 1950s Hollywood and the Rebirth of Low-Budget Cinema* (New Brunswick, New Jersey: Rutgers University Press, 2012).

Davis, Edward B., 'Fundamentalist cartoons, modernist pamphlets, and the religious image of science in the Scopes era', in Charles L. Cohen and Paul S. Boyer (eds.), *Religion and the Culture of Print in Modern America* (Madison: University of Wisconsin Press, 2008), pp. 175–98.

Davis, Jim, 'Imperial transgressions: the ideology of Drury Lane pantomime in the late nineteenth century', *New Theatre Quarterly*, 12:46 (1996), pp. 147–55.

Dawson, Gowan, *Darwin, Literature and Victorian Respectability* (Cambridge: Cambridge University Press, 2007).

Degler, Carl N., *In Search of Human Nature: The Decline and Revival of Darwinism in American Social Thought* (New York: Oxford University Press, 1991).

Desmond, Adrian, 'Artisan resistance and evolution in Britain, 1819–1848', *Osiris*, 3 (1987), pp. 77–110.

——*Huxley: From Devil's Disciple to Evolution's High Priest* (Reading: Addison Wesley, 1997).

—— 'Richard Owen's reaction to transmutation in the 1830s', *British Journal for the History of Science*, 18:1 (1985), pp. 25–50.

—— 'Richard Owen's response to Robert Edmond Grant', *Isis*, 70:2 (1979), pp. 224–34.

—— and James Moore, *Darwin's Sacred Cause: Race, Slavery and the Quest for Human Origins* (London: Allen Lane, 2009).

Dickins, Bruce, 'Samuel Page Widnall and his press at Grantchester, 1871–1892', *Transactions of the Cambridge Bibliographical Society*, 2:5 (1958), pp. 366–72.

Disraeli, Benjamin, *Tancred, or the New Crusade* (London: Henry Colburn, 1847).

Dixon, Bryony, 'The ancient world on silent film: the view from the archive', in Pantelis Michelakis and Maria Wyke (eds.), *The Ancient World in Silent Cinema* (Cambridge: Cambridge University Press, 2013), pp. 27–36.

Dupree, A. Hunter, 'Introduction', in A. Hunter Dupree (ed.), *Darwiniana: Essays and Reviews Pertaining to Darwinism by Asa Gray* (Cambridge, Massachusetts: Harvard University Press, 1963), pp. ix–xxiii.

Durbach, Nadja, *Spectacle of Deformity: Freak Shows and Modern British Culture* (Oakland: University of California Press, 2009).

Dyson, Will, *Kultur Cartoons* (London: Stanley and Paul, 1915).

Edwards, Rebecca, 'Politics as social history: political cartoons in the Gilded Age', *OAH Magazine of History*, 13:4 (1999), pp. 11–15.

'Edward Tennyson Reed', in L.G. Wickham Legg (ed.), *The Dictionary of National Biography, 1931–1940* (London: Oxford University Press, 1949), p. 730.

Eliot, Sir Charles, *Report by Her Majesty's Commissioner on the East Africa Protectorate* (London: His Majesty's Stationery Office, 1901).

Ellegård, Alvar, *Darwin and the General Reader* (Chicago: University of Chicago Press, 1990).

—— 'Public opinion and the press: reactions to Darwinism', *Journal of the History of Ideas*, 19:3 (1958), pp. 379–87.

—— 'The readership of the periodical press in mid-Victorian Britain', special issue, *Acta Universitatis Gothoburgensis*, 63:3 (1957).

Fara, Patricia, 'Pictures of Charles Darwin', *Endeavour*, 24:4 (2000), pp. 143–4.

Fichman, Martin, *Evolutionary Theory and Victorian Culture* (Amherst, New York: Humanity Books, 2002).

Fielding, William J., *The Caveman Within Us* (New York: Dutton, 1922).

Figuier, Louis, *Primitive Man* (London: Chapman and Hall, 1870).

Finkel, Irving, *The Ark before Noah: Decoding the Story of the Flood* (London: Hodder and Stoughton, 2014).

Foot, M.R.D. (ed.), *The Gladstone Diaries*, vol. VIII (Oxford: Clarendon Press, 1968).

Fretz, Eric, 'P.T. Barnum's theatrical selfhood and the nineteenth-century culture of exhibition', in Rosemarie Garland Thomson (ed.), *Freakery: Cultural Spectacles of the Extraordinary Body* (New York: New York University Press, 1996), pp. 97–107.

Further Correspondence Relating to the Island of Tristan da Cunha (London: His Majesty's Stationery Office, 1903).

Gaige, Roscoe C., *The Missing Link* (New York: Brooks & Denton, 1904).

Gamble, Clive, 'Foreword', in Stephanie Moser, *Ancestral Images: The Iconography of Human Origins* (Ithaca: Cornell University Press, 1998), pp. ix–xxiv.

Gifford, Denis, 'Fitz: the old man of the screen', in Charles Barr (ed.), *All Our Yesterdays* (London: British Film Institute, 1986), p. 314–20.

Goodall, Jane R., *Performance and Evolution in the Age of Darwin: Out of the Natural Order* (London: Routledge, 2002).

Gordon, Rae Beth, 'Natural rhythm: la Parisienne dances with Darwin, 1875–1910', *Modernism/Modernity*, 10:4 (2003), pp. 617–56.

Gray, Beverley, *Roger Corman: An Unauthorized Biography of the Godfather of Indie Filmmaking* (Los Angeles: Renaissance Books, 2000).

Griest, Guinevere L., 'A Victorian leviathan: Mudie's Select Library', *Nineteenth-Century Fiction*, 20:2 (1965), pp. 103–26.

Groves, Colin, 'A history of gorilla taxonomy', in Andrea B. Taylor and Michele L. Goldsmith (eds.), *Gorilla Biology: A Multidisciplinary Perspective* (Cambridge: Cambridge University Press, 2002), pp. 15–34.

Hall, John Elihu, *The Philadelphia Souvenir* (Philadelphia: William Brown, 1826).
Hanna, Bill, with Tom Ito, *A Cast of Friends* (Dallas: Taylor Publishing, 1996).
Harding, James, *George Robey and the Music Hall* (London: Hodder and Stoughton, 1990).
Heller, Michael, 'Work, income and stability: the late Victorian and Edwardian London male clerk revisited', *Business History*, 50:3 (2008), pp. 253–71.
Henisch, Heinz K., and Bridget A. Henisch, *Positive Pleasures: Early Photography and Humor* (University Park: Pennsylvania State University Press, 1998).
'Henry James Byron', in Leslie Stephen and Sidney Lee (eds.), *Dictionary of National Biography*, vol. III (London: Smith Elder, 1908), pp. 607–9.
Hepworth, Cecil, *Came the Dawn: Memories of a Film Pioneer* (London: Phoenix House, 1951).
Hesketh, Ian, *Of Apes and Ancestors: Evolution, Christianity and the Oxford Debate* (Toronto: University of Toronto Press, 2009).
Higson, Andrew, 'Introduction', in Andrew Higson (ed.), *Young and Innocent? The Cinema in Britain 1896–1930* (Exeter: University of Exeter Press, 2002), pp. 1–12.
Hodgson, Amanda, 'Defining the species: apes, savages and humans in scientific and literary writing of the 1860s', *Journal of Victorian Culture*, 4:2 (1999), pp. 228–51.
Hoffer, Tom W., 'From comic strips to animation: some perspectives on Winsor McCay', *Journal of the University Film Association*, 28:2 (1976), pp. 23–32.
Homchick, J., 'Objects and objectivity: the evolution controversy at the American Museum of Natural History, 1915–1928', *Science & Education*, 19:4–5 (2010), pp. 485–503.
Hood, Basil, and Arthur Sullivan, *The Rose of Persia, or the Story-teller and the Slave* (London: Chappell, 1899).
Hutchinson, G., 'Hindu pilgrimages in India', *British Medical Journal* (20 October 1894), p. 907.
Hutchinson, H.N., *Extinct Monsters* (London: Chapman and Hall, 1897).
—— *Primeval Scenes: Being Some Comic Aspects of Life in Prehistoric Times* (London: Lamley, 1899).
Huxley, Leonard (ed.), *The Life and Letters of Thomas Henry Huxley* (Cambridge: Cambridge University Press, 2012).
Huxley, Thomas Henry, *Evidence as to Man's Place in Nature* (New York: Appleton, 1863).
—— *Lay Sermons, Addresses and Reviews* (London: Macmillan, 1870).
Isaacs, George, *Rhyme and Prose: A Burlesque and its History* (Melbourne: Clarson, Shallard, 1865).
Janson, H.W., *Apes and Ape Lore in the Middle Ages and the Renaissance* (London: Warburg Institute, 1952).
Keith, Sir Arthur, *An Autobiography* (New York: Philosophical Society, 1950).
Kent, Christopher, 'War cartooned/cartoon war: Matt Morgan and the American Civil War in *Fun* and *Frank Leslie's Illustrated Newspaper*', *Victorian Periodicals Review*, 36:2 (2003), pp. 153–81.
Kift, Dagmar, *The Victorian Music Hall: Culture, Class and Conflict* (Cambridge: Cambridge University Press, 1996).

King, Clive, *Stig of the Dump* (London: Puffin, 2003).
King, Peter (ed.), *Scott's Last Journey* (London: Harper Collins, 1999).
Kingsley, Charles, *Charles Kingsley: His Letters and Memories of his Life, Edited by his Wife* (New York: Scribner Armstrong, 1877).
—— *The Water Babies* (London: Macmillan, 1910).
Kipling, Rudyard, *Rudyard Kipling's Verse: Inclusive Edition* (New York: Doubleday, 1918).
Kjærgaard, Peter C., '"Hurrah for the missing link!": a history of apes, ancestors and a crucial piece of evidence', *Notes and Records: The Royal Society Journal of the History of Science*, 65:1 (2011), pp. 83–98.
Klindt-Jensen, Ole, 'Dating the earliest Iron Age in Scandinavia', in John D. Evans, Barry Cunliffe and Colin Renfrew (eds.), *Antiquity and Man: Essays in Honour of Glyn Daniel* (London: Thames and Hudson, 1981), pp. 25–7.
Klossner, Michael, *Prehistoric Humans in Film and Television: 581 Dramas, Comedies and Documentaries, 1905–2004* (Jefferson, North Carolina: McFarland, 2006).
Lancaster, Osbert, *The Littlehampton Saga* (London: Methuen, 1984).
Landseer, Thomas, *Monkeyana, or Men in Miniature* (London: Moon, Boys and Graves, 1827).
Laurent, John, 'Science, society and politics in late nineteenth-century England: a further look at mechanics' institutes', *Social Studies of Science*, 14:4 (1984), pp. 585–619.
Leacock, Stephen, *Frenzied Fiction* (London: Bodley Head, 1917).
Lenburg, Jeff, *William Hanna and Joseph Barbera: The Sultans of Saturday Morning* (New York: Chelsea House, 2011).
Leverenz, David, 'The last real man: from Natty Bumpo to Batman', *American Literary History*, 3:4 (1991), pp. 753–81.
Levine, Philippa, *The Amateur and the Professional: Antiquarians, Historians and Archaeologists in Victorian England, 1838–1886* (Cambridge: Cambridge University Press, 1986).
Lewis, Roy, *The Evolution Man: or How I Ate My Father* (New York: Pantheon Books, 1960).
Lightman, Bernard, *Victorian Popularizers of Science: Designing Nature for New Audiences* (Chicago: University of Chicago Press, 2007).
Lively, James K., 'Propaganda techniques of Civil War cartoonists', *Public Opinion Quarterly*, 6:1 (1942), pp. 99–106.
Lomas, Sophie C., *Festival of Empire: Souvenir of the Pageant of London* (London: Bemrose and Sons, 1911).
Long, Helen C., *The Edwardian House: The Middle Class Home in Britain, 1880–1914* (Manchester: Manchester University Press, 1993).
Lubbock, John, *Pre-Historic Times* (London: Williams and Norgate, 1865).
McCook, Stuart, '"It may be truth, but it is not evidence": Paul Du Chaillu and the legitimation of evidence in the field sciences', *Osiris*, 11 (1996), pp. 177–97.
McLennan, John F., *Primitive Marriage* (Edinburgh: Adam and Charles Black, 1865).
Meade, Marion, *Buster Keaton: Cut to the Chase* (New York: Harper Collins, 1995).
Methley, Noel T., *The Life-Boat and its Story* (London: Sidgwick and Jackson, 1912).
Michelakis, Pantelis, and Wyke, Maria, 'Introduction: silent cinema, antiquity and "the

exhaustless urn of time"', in Pantelis Michelakis and Maria Wyke (eds.), *The Ancient World in Silent Cinema* (Cambridge: Cambridge University Press, 2013), pp. 1–24.

Michelakis, Pantelis, and Maria Wyke (eds.), *The Ancient World in Silent Cinema* (Cambridge: Cambridge University Press, 2013).

Miller, Henry J., 'John Leech and the shaping of the Victorian cartoon: the context of respectability', *Victorian Periodicals Review*, 42:3 (2009), pp. 267–91.

Mitchell, W.J.T., *The Last Dinosaur Book: The Life and Times of a Cultural Icon* (Chicago: University of Chicago Press, 1998).

Moews, Daniel, *Keaton: The Silent Features Close Up* (Berkeley: University of California Press, 1977).

Moore, James R., *The Post-Darwinian Controversies: A Study of the Protestant Struggle to Come to Terms with Darwin in Great Britain and America, 1870–1900* (Cambridge: Cambridge University Press, 1979).

Moran, Jeffrey P., *American Genesis: The Antievolution Controversies from Scopes to Creation Science* (New York: Oxford University Press, 2012).

Morley, Henry, *Memoirs of Bartholomew Fair* (London: Chapman and Hall, 1859).

Moser, Stephanie, *Ancestral Images: The Iconography of Human Origins* (Ithaca: Cornell University Press, 1998).

—— and Clive Gamble, 'Revolutionary images: the iconic vocabulary for representing human antiquity', in Brian Leigh Molyneux (ed.), *The Cultural Life of Images: Visual Representation in Archaeology* (London: Routledge, 1997), pp. 184–204.

Nielsen, Wendy C., *Women Warriors in Romantic Drama* (Wilmington: University of Delaware Press, 2013).

Noakes, Richard, '*Punch* and mid-Victorian comic journalism', in Geoffrey Cantor et al., *Science in the Nineteenth-Century Periodical: Reading the 'Magazine of Nature'* (Cambridge: Cambridge University Press, 2004), pp. 91–122.

—— 'Science in mid-Victorian *Punch*', *Endeavour*, 26:3 (2002), pp. 92–144.

Numbers, Ronald L., *The Creationists: From Scientific Creationism to Intelligent Design* (Cambridge, Massachusetts: Harvard University Press, 2006).

—— *Darwinism Comes to America* (Cambridge, Massachusetts: Harvard University Press, 1998).

Opper, Frederick Burr, *Our Antediluvian Ancestors* (London: Pearson, 1903).

Overy, Richard, *The Morbid Age: Britain between the Wars* (London: Allen Lane, 2009).

Paradis, James G., 'Satire and science in Victorian culture', in Bernard Lightman (ed.), *Victorian Science in Context* (Chicago: University of Chicago Press, 1997), pp. 143–75.

Peacock, Shane, *The Great Farini: The High-Wire Life of William Hunt* (Toronto: Viking, 1995).

Peacock, Thomas Love, *Melincourt: or Sir Oran Haut-Ton* (London: Macmillan, 1896).

Pearsall, Ronald, *Victorian Sheet Music Covers* (Detroit: Gale, 1972).

Pearson, Richard, 'Primitive modernity: H.G. Wells and the prehistoric man of the 1890s', *Yearbook of English Studies*, 37:1 (2007), pp. 58–74.

'Piebald', in *The Oxford English Dictionary*, vol. VII (Oxford: Clarendon Press, 1978), pp. 834–5.

Pina, Marco, and Nathalie Gontier (eds.), *The Evolution of Social Communication in Primates* (New York: Springer, 2014).

Planché, J.R., *The Recollections and Reflections of J.R. Planché*, vol. II (London: Tinsley Brothers, 1872).

Potter, Simon, 'Communication and integration: the British and Dominions press and the British world, c. 1876–1914', *Journal of Imperial and Commonwealth History*, 31:2 (2003), pp. 190–206.

Proctor, Tammy M., 'On my honour: Guides and Scouts in interwar Britain', *Transactions of the American Philosophical Society*, 92:2 (2002), pp. 1–180.

Quammen, David, *The Boilerplate Rhino: Nature in the Eye of the Beholder* (New York: Oxford University Press, 2000).

Quayle, Eric, *Ballantyne the Brave: a Victorian Writer and his Family* (London: Hart-Davis, 1967).

Qureshi, Saddiah, *Peoples on Parade: Exhibitions, Empire, and Anthropology in Nineteenth-Century Britain* (Chicago: University of Chicago Press, 2011).

Reed, E.T., *Mr Punch's Animal Land* (London: Bradbury and Agnew, 1898).

—— *Prehistoric Peeps* (London: Bradbury and Agnew, 1896).

—— *Prehistoric Peeps* (London: Bradbury and Agnew, 1902).

—— *Punch's Holiday Book* (London: Bradbury and Agnew, 1901).

Reel, Monte, *Between Man and Beast: An Unlikely Explorer, the Evolution Debates, and the African Adventure that Took the Victorian World by Storm* (New York: Doubleday, 2013).

Rhys Davids, C.A.F, 'Il suicidio nel diritto e nella vita sociale', *Mind* (October 1907), p. 617.

Richards, Jeffrey, *The Ancient World on the Victorian and Edwardian Stage* (London: Palgrave Macmillan, 2009).

—— *The Golden Age of Pantomime: Slapstick, Spectacle and Subversion in Victorian England* (London: I.B. Taurus, 2015).

—— *Hollywood's Ancient Worlds* (London: Continuum, 2008).

Richards, Thomas, *The Commodity Culture of Victorian England: Advertising and Spectacle, 1851–1914* (Stanford: Stanford University Press, 1990).

Rightmire, G. Philip, 'Human evolution in the Middle Pleistocene: the role of *Homo Heidelbergensis*', *Evolutionary Anthropology*, 6:6 (1998), pp. 218–27.

Robey, George, *Looking Back on Life* (London: Constable, 1933).

—— *My Life up till Now* (London: Greening, 1908).

Robinson, David, *From Peep Show to Palace: The Birth of American Film* (New York: Columbia University Press, 1996).

Rodmell, Paul, *Opera in the British Isles, 1875–1918* (Aldershot: Ashgate, 2013).

Rothfels, Nigel, 'Aztecs, aborigines, and ape-people: science and freaks in Germany, 1850–1900', in Rosemarie Garland Thomson (ed.), *Freakery: Cultural Spectacles of the Extraordinary Body* (New York: New York University Press, 1996), pp. 158–72.

Ruddick, Nicholas, 'Courtship with a club: wife-capture in prehistoric fiction, 1865–1914', *Yearbook of English Studies*, 37:2 (2007), pp. 45–63.

—— *The Fire in the Stone: Prehistoric Fiction from Charles Darwin to Jean M. Auel* (Middletown: Wesleyan University Press, 2009).

Rudwick, Martin J.S., *Scenes from Deep Time: Early Pictorial Representations of the Prehistoric World* (Chicago: University of Chicago Press, 1992).

—— *Worlds before Adam: The Reconstruction of Geohistory in the Age of Reform* (Chicago: University of Chicago Press, 2009).

Rupke, Nicolaas A., *Richard Owen: Victorian Naturalist* (New Haven: Yale University Press, 1994).

Sabin, Roger, 'Ally Sloper on stage', *European Comic Art*, 2:2 (2009), pp. 205–25.

'Samuel Wilberforce', in Sidney Lee (ed.), *Dictionary of National Biography*, vol. XXI (London: Smith Elder, 1909), pp. 204–8.

Scholnick, Robert J., 'The fate of humor in a time of civil and cold war: *Vanity Fair* and race', *Studies in American Humor*, 3:10 (2003), pp. 21–42.

Scully, Richard, 'A comic empire: the global expansion of *Punch* as a model publication, 1841–1936', *International Journal of Comic Art*, 15:2 (2013), pp. 6–35.

—— '"The epitheatrical cartoonist": Matthew Somerville Morgan and the world of theatre, art and journalism in Victorian London', *Journal of Victorian Culture*, 16:3 (2011), pp. 363–84.

Secord, Anne, 'Science in the pub: artisan botanists in early nineteenth-century Lancashire', *History of Science*, 32 (1994), pp. 269–315.

Secord, James A., 'Monsters at the Crystal Palace', in Soraya de Chadarevian and Nick Hopwood (eds.), *Models: The Third Dimension in Science* (Stanford: Stanford University Press, 2004), pp. 138–69.

—— *Victorian Sensation: The Extraordinary Publication, Reception and Secret Authorship of 'Vestiges of the Natural History of Creation'* (Chicago: University of Chicago Press, 2000).

Sellers, Charles Coleman, 'Peale's Museum and the "museum idea"' *Proceedings of the American Philosophical Society*, 124:1 (1980), pp. 25–34.

Semonin, Paul, *American Monster: How the Nation's First Prehistoric Creature Became a Symbol of National Identity* (New York: New York University Press, 2000).

—— 'Monsters in the market place: the exhibition of human oddities in early modern England', in Rosemarie Garland Thomson (ed.), *Freakery: Cultural Spectacles of the Extraordinary Body* (New York: New York University Press, 1996), pp. 69–81.

Shipman, Pat, *The Man who Found the Missing Link* (London: Simon & Schuster, 2001).

Skelton, Matthew, 'The paratext of everything: constructing and marketing H.G. Wells's *The Outline of History*', *Book History*, 4 (2001), pp. 237–75.

Skretvedt, Randy, and Jordan R. Young, *Laurel and Hardy: The Magic behind the Movies* (Beverly Hills: Moonstone Press, 1987).

Smith, David C., *H.G. Wells: Desperately Mortal, a Biography* (New Haven: Yale University Press, 1986).

Snigurowicz, Diana, 'Sex, simians and spectacle in nineteenth century France: or how to tell a man from a monkey', *Canadian Journal of History*, 34 (1999), pp. 51–81.

Sommer, Marianne, 'Mirror, mirror on the wall: Neanderthal as image and "distortion" in early 20th-century French science and press', *Social Studies of Science*, 36:2 (2006), pp. 207–40.

Stevenson, Robert Louis, *Treasure Island* (London: Cassell, 1883).

Stott, Rebecca, *Darwin's Ghosts: In Search of the First Evolutionists* (London: Bloomsbury, 2012).

Strange, Carolyn, 'Reconsidering the "tragic" Scott expedition: cheerful masculine home-making in Antarctica, 1910–1913', *Journal of Social History*, 46:1 (2012), pp. 66–88.

Taylor, Andrea B., and Michele L. Goldsmith (eds.), *Gorilla Biology: A Multidisciplinary Perspective* (Cambridge: Cambridge University Press, 2002).

Toole, J.L., *Reminiscences of J.L. Toole Related by Himself and Chronicled by Joseph Hatton* (London: George Routledge and Sons, 1892).

Trinkhaus, Erik, and Pat Shipman, *The Neanderthals: Changing the Image of Mankind* (New York: Alfred A. Knopf, 1993).

True, John Preston, *The Iron Star* (London: Gay and Bird, 1899).

Van Hare, W.G., *Fifty Years of a Showman's Life* (London: W. H. Allen, 1888).

Van Riper, A. Bowdoin, *Men among the Mammoths: Victorian Science and the Discovery of Human Prehistory* (Chicago: University of Chicago Press, 1993).

Vénus et Caïn: Figures de la préhistoire, 1830–1930 (Bordeaux: Éditions de la réunion des musées nationaux, 2003).

Walker, Thomas 'Whimsical', *From Sawdust to Windsor Castle* (London: Stanley Paul, 1922).

Walsh, John Evangelist, *Unravelling Piltdown* (New York: Random House, 1996).

Ward, Richard Lewis, *A History of the Hal Roach Studios* (Carbondale: Southern Illinois University Press, 2005).

Waterloo, Stanley, *The Story of Ab* (Chicago: Way and Williams, 1897).

Wells, H.G., *The Outline of History* (London: Cassell & Co., 1920).

—— *Tales of Space and Time* (London: Macmillan, 1920).

White, Paul, *Thomas Huxley: Making the 'Man of Science'* (Cambridge: Cambridge University Press, 2003).

Widnall, Samuel Page, *A Mystery of Sixty Centuries, or a Modern St. George and the Dragon* (Grantchester: S.P. Widnall, 1889).

Wilson, Albert, *The Prime Minister of Mirth: The Biography of Sir George Robey* (London: Odhams, 1956).

Wood, J. Hickory, *Dan Leno* (London: Methuen, 1905).

Wood, Lawson, *Prehistoric Proverbs* (London: Collier, 1907).

Wyke, Maria, 'Silent laughter and the counter-historical: Buster Keaton's "Three Ages"', in Pantelis Michelakis and Maria Wyke (eds.), *The Ancient World in Silent Cinema* (Cambridge: Cambridge University Press, 2013), pp. 275–96.

Yochelson, Ellis L., 'Mr Peale and his mammoth museum', *Proceedings of the American Philosophical Society*, 136:4 (1992), pp. 487–506.

Index

Note: literary works are found under authors' names
Note: italicised page numbers refer to illustrations

acrobats and gorilla acrobats 15, 16, 23, 30, 40–1, 43, 46–7, 60, 61–2, 63, 64–7, 73, 83, 118–19, 177, 199, 201, *42*
Addison, Henry *Blue-Faced Baboon, The* 38, 80
advertisements, cave men and prehistory used in 11, 20–1, 34, 41, 60–2, 66–7, 71–2, 81, 98, 108, 118–19, 124, 129, 147–8, 150, 153, 155, 161–2, 171, 174, 188–9, *126, 128, 175*
Africa 6, 15, 16, 20, 30–5, 38, 43, 46, 48, 50, 61–3, 66, 71–3, 84–5, 92–3, 101, 103, 115, 129, 137, 150, 154, 199–200, *94, 103*
American Museum of Natural History 134, 150, 176
archaeology 7, 13, 14, 17, 34, 58, 69, 85, 97, 105
Auel, Jean M. *Clan of the Cave Bear* 191
Australia 2, 19, 40, 46, 97, 101, 119, 121, 129, 136, 148, 150, 151, 160–1, 164, 174, 176, *120, 175*
Avon Tyres 127, 147, *128*

Ballantyne, R.M. *Gorilla Hunters, The* 48–9
Barbera, Joseph 188–9
Barnum, Phineas Taylor, 'P.T.' 20–2, 24, 34–6, 44, 62, 67, 69, 71, 81, 89, 139
Bartholomew Fair 15, 20
Battel, Andrew 16, 61
Blondin, Charles 40–1, 43, 46, 61
Boadicea Unearthed 104–5, 116
Boers, comparisons to prehistoric humans 62, 101
Boitard, Pierre *Paris avant les hommes* 85
Boucicault, Dion 34, 43
Boy Scouts, use of cave men and prehistory in activities of 148, 153, 172–4, *173*

British Association for the Advancement of Science 17, 31, 33, 40, 50, 69–70, 88
British history, popular conceptions of 14, 83–4, 98, 122, 127, 129, 200
British Museum 12, 22, 35, 96
Brixham cave 13
Buckland, William 12
Burma 68
Burroughs, Edgar Rice *Tarzan of the Apes* 137, 150, 173
Byron, Henry J. 37, 38, 43, 44, 46

Cambridge, University of 57, 104
Campbell-Bannerman, Sir Henry 95, 123, *122*
carnivals *see* pageants, parades and carnivals
cartoon, concept of 19
cartoonists 90, 96, 107, 137, 150, 182, 190
  see also individual cartoonists
Chamberlain, Sir Joseph 89, 95, 122
Chaplin, Charlie *His Prehistoric Past* 162
Chesterton, Gilbert Keith 'G.K.' 177
Chevalier, Albert 123
circuses and menageries 46, 47, 60, 67, 71, 81, 177
Cold War 183–4
Coleridge, Samuel Taylor 15
Collins, Wilkie *Moonstone, The* 17
colloquial speech, allusions to cave men and prehistory in 11, 37, 40, 63, 72, 97–8, 115, 156, 199
colonial society, prehistoric allusions used in 6, 11, 62, 83, 98, 100–4, 121, 134, 147, *94, 102, 103*
Concanen, Alfred 39, 40
Crystal Palace, The 16, 22, 30, 40, 46, 60, 65, 129

Darwin, Charles 4–5, 11, 25, 31–3, 36, 46–50, 57–60, 63–7, 70–1, 81, 89, 92, 127, 138, 176–7, 183, 192, 198–9, *82*
  Descent of Man, and Selection in Relation to Sex, The 57
  Expression of the Emotions in Man and Animals 57
  On the Origin of Species 4–5, 25, 31–3, 57
Dickens, Charles 47, 124
dinosaurs, discovery of 12, 60–1, 88–9, 134, 138, 150, 202
  models of 22, 60
Disraeli, Benjamin 18, 49–50, 63, 66
  Tancred: or the New Crusade 18
Du Chaillu, Paul 33–43, 46, 47, 48, 49, 61, 62, 73
Dubois, Eugène 80–1, *82*
Dundreary, Lord 33, 48, 59–60
Dyson, Will 151–2, 156

Egerton, Sir Philip 'Monkeyana' 35
evolution, popular understanding of 4–5, 10–17, 20–1, 24–5, 30–5, 36, 49–50, 57–60, 63, 64, 66–9, 72, 73, 81, 83, 88–92, 97, 118, 127, 132, 139, 176–7, 199–200, *93*

Farini, Guillermo 61–73, 84, 139, *70*
Figuier, Louis 85, 98
films and filmmakers 2–3, 135–40, 148–50, 161–2, 174–92, 179–86, 202, *149*
First World War, prehistory as a way of conceptualising 150–60, 163, 164–5, *157, 158, 159*
Flintstones, The 4, 187–9, 192, 198, 201
freaks and freak shows 20–2, 35, 44, 67–73, 81, 83–4, 87, 199
  'What Is It?' 21, 34, 36, 44, 62, 67, 71, 81
French people, depicted as cave men 92–5, *94*
frontier, influence on American idea of cave men 6, 11, 133–4, 136–9, 165

games *see* sport and games, prehistoric visions of
George, William 122, 147
Germans, compared to prehistoric hominids 150–3, 157
Gladstone, William Ewart 13, 34–5, 57–8, 66, 89
gorilla acrobats *see* acrobats
'gorilla quadrille, The' (dance) 39, 46, 60

gorillas 4, 15–16, 30–50, 59–68, 72–3, 80–1, 83–5, 87, 90, 134, 139, 148, 151, 152, 189, 199, 201
Gray, Asa 33
Gray, John Edward 22, 35
Griffith, D.W. 138–9, 162, 178

Hanna, William 188–9
Hawkins, Benjamin Waterhouse 22, 40, 49, 60, 88
Hepworth, Cecil 135–6
Hicks, Seymour 148, 162, 180, *149*
history *see* British history, popular conceptions of
humbug, concept of 21–2, 25, 35, 46, 48, 58, 68–9, 71, 73, 89, 139, 199
Hutchinson, Reverend Henry Neville
  Primeval Scenes: Being Some Comic Aspects of Life in Prehistoric Times 108
Huxley, Thomas Henry 31–3, 49, 57, 60, 67, 72, 80, 81, 88, 92, 106, *93*
  Evidence as to Man's Place in Nature 49

images, importance and influence of 4, 18, 176–7
indigenous peoples, depictions of 20, 35, 69, 81, 84, 98, 103–4, 121, 134, 138, 139, 192, 199, *70, 82, 94, 103*
Irish, compared to prehistoric hominids 47–8, 62, 72

Jocko (play) 15–16, 41

Keaton, Buster *Three Ages* 178–9, 181
King Mugslot (film) 148–50, 162, 180, *149*
King, Clive *Stig of the Dump* 187
Kingsley, Charles 33, 48–9
  Water Babies, The 48–9
Kipling, Rudyard 105–6, 124, 127, 150, 173, 201
  'In the Neolithic Age' 105
  'story of Ung, The' 105
Kirkdale cave 12
Kotaki 64–7
Krao 67–73, 87, *70*

Lancaster, Osbert 'Draynefleet revealed' 186
Landseer, Thomas *Monkeyana* 15–16
Lankester, Professor Ray 170, *171*
Lauri, George 119–20, *120*
Leacock, Stephen 'cave-man as he is, The' 165
Leech, Hervey 21
Leno, Dan 65, 95
Léotard, Jules 41, 43, 47

Lewis, Roy
What We Did to Father 187
Lloyd, Marie 119, 131
'When I take my morning promenade' (song) 131
London, Jack *Before Adam* 137
Lubbock, Sir John, coins term cave man 7, 58
Lyell, Sir Charles 15
Lyme Regis 12

magazines and newspapers, importance of 19–20, 34, 64, 90
mammoths and mastodons, 6, 11, 14, 22, 87, 89, 155, 186, 188
mass culture, concept of 18
menageries *see* circuses and menageries
military, adoption of prehistoric imagery by 101–3, 129, *102*
  see also First World War, prehistory as a way of conceptualising
minstrels and minstrel shows 23, 47–8, 65, 66, 93, 121
missing link, concept of 4–5, 83–4, 139
  supposed examples of 48, 58–73, 80–1 *82*
models and waxworks 22, 44, 60–1, 71, 79n.73
Molesey Boat Club 98–9
monkeys 14–16, 21
Monkeyana, *Punch* 35–6
'Mr Gorilla' cartoon, play and song 36–41, 43–6, 48, *45*
Mudie's Select Library 17, 31, 35
Murchison, Sir Roderick 34–5, 49
Murray, John 35
music hall 22–5, 36, 40–4, 47, 49, 50, 59, 61, 63–6, 69, 71, 83, 91, 93, 95, 99, 109, 116–19, 123, 129, 135, 147, 149, 180

New Yorker magazine 181, 202
New Zealand 98, 101, 127, 129, 133, 136, 150, 155, 161, 174, 176, 190
newspapers *see* magazines and newspapers

On the Origin of Species see Darwin, Charles
Opper, Frederick Burr 90–1, 133–4, 136, 181–2, 192, 201
orang-utans 15, 21, 49, 85, 139
Owen, Sir Richard 12, 15–16, 22, 24, 30–5, 40, 49, 62, 80, 81
Oxford debate (1861) 31–2, 36

pageants, parades and carnivals 16, 37, 124, 129, 131, 135, 139, 148, 154, 164, 171–4, 179, 186, 200

palaeontology 7, 35, 60, 64, 73, 80, 86, 88, 134, 186–7, 192, *82*
pantomimes *see* sketches, plays, revues and pantomimes
parades *see* pageants, parades and carnivals
Paul, Howard 38–9, 46, 59, 63, 64
'parents of Adam and Eve: according to Darwin, The' (song) 59
Peacock, Thomas Love *Melincourt: or Sir Oran Haut-Ton* 15
Peale, Charles Willson 11, 22
Pepper, John Henry 16
Piltdown Man 139
Planché, James 37, 59
plays *see* sketches, plays, revues and pantomimes
politics, references to cave men and prehistory in 15, 58, 63, 66, 72, 89, 95, 123, 147, 186, *122*
Pongo the gorilla 61–9, 73, 87
Pongo Redivivus 64, 66–7
prehistory and science, popular interest in 4–5, 6, 10–18, 15, 31, 34–5, 60, 69–70, 80–1, 85, 90–1, 106, 108, 116, 151–2, 155, 162–3, 183, 186–7, 191–2, *82*
'prehistoric cakewalk, The' (song) 122
Prehistoric peeps cartoons 5–6, 87–9, 91–8, *93, 94, 99*
  allusions to 97–8, 101, 103, 119, 121, 127–9, 133, 147, 155, 170, 186
  imitative cartoons 96–8, 132–4, 122 *103, 122, 157, 158*
  recreating scenes from 98–101, *102*
'Prehistoric zig zags' (song) 147
publishing industry 17–18, 24, 35
Punch (magazine) 5, 19–22, 32, 35–6, 38–41, 44–6, 48, 50, 59, 62–3, 80, 85, 87–101, 103, 105, 115, 117–18, 121, 123, 129, 132–5, 151, 156, 181, 190, 198–9, 202, *93, 94, 99*

race, concept and depictions of 6, 20, 32–4, 36, 40, 47–8, 62, 72–3, 83–6, 91–3, 102–4, 108, 115, 134, 138–9, 148, 200
Reed, Edward Tennyson 'E.T.' 5–6, 87–109, 115–18, 121–47, 129, 131–6, 148–9, 153, 155–8, 160–1, 163, 165, 170, 174, 178, 181, 183–4, 187–90, 192, 198–202, *93, 94, 99, 22*
religious responses to prehistory and evolution 12–14, 31–3, 41–3, 57–8, 176–7, *93*
revues *see* sketches, plays, revues and pantomimes

Robey, George 1–4, 116–21, 129, 131–2, 136, 138, 148, 150, 153, 162–5, 178, 179–81, 201, *3*
  Prehistoric Man, The (film) 179–80
  'prehistoric man, The' (sketch) 116–18, 129, 131, 162–4, *3*
Rose of Persia (Savoy opera) 105
royal family 22, 63, 71, 104, 129, 178
Royal Geographical Society 34, 35

satire, concept of 19, 33, 65, 90, 198–200, 202
Scopes Monkey Trial 177
Scott, Captain Robert Falcon 1–2
Shannon, Charles 85, *86*
Shentini, Herr 41
sketches, plays, revues and pantomimes 34, 36–7, 43, 44, 47, 59–60, 65–7, 105, 119, 121, 123–4, 134, 147–50, 162–4, 174, 179, 199, *125*
slang *see* colloquial speech, allusions to cave men and prehistory in
slavery *see* race, concept of
Smith, George *Epic of Gilgamesh* 57–8
Spencer, Herbert and Social Darwinism 60
sport and games, prehistoric visions of 95, 99–100, 101, 120–1, 129, 131, 153–4, 172, 174, 178, *99, 102, 120*
Spurgeon, Charles 41–3, 46
Stanley, Henry Morton 59, 72
Stanworth's Umbrellas 124, 126–7, 161, 171, *126*
Sterling, Pedro 66–7
Stevenson, Robert Louis *Treasure Island* 72
Stonehenge 97, 99, 122, 127, 131, 177, 180, 200

television series 187–9, 191–2
theatrical industry 24, 40–4
Times Literary Supplement 198–9
Traill, Henry Duff 84–5, 105
  'day with primeval man, A' 84–5, *86*

Tree, Herbert Beerbohm, *Tempest, The* 121
trench newspapers 155–60, *157, 158, 159*
True, John Preston *Iron Star, The* 107
Turks, compared to prehistoric hominids 62, 65, 66

United States of America 6, 7, 11, 19–20, 21, 22, 23, 32–4, 40, 43–5, 47, 59, 60, 61, 71, 90, 98, 115, 122, 133–4, 136, 138, 147, 150, 152, 161, 162, 165, 174, 176, 178, 183–4, 188, 190, 192, 200, 202, *45*
  cartoons 44–5, 90–1, 98, 133–4, *45*
  cave men in 60, 136–9, 165
  Christian fundamentalism in *see* religious responses to prehistory and evolution
  evolutionary ideas in 32–4

Van Hare, G.A. 46
variety entertainment 23, 83, 117, 135, 179, 180
Vestiges of the Natural History of Creation 17–18, 21

Waterloo, Stanley *Story of Ab, The* 106–7
Welch, Raquel 189
Wells, Herbert George 'H.G.' 106–7, 170, *171*
  Outline of History, The 170
  'story of the Stone Age, A' 106–7
Westminster Aquarium 61–2, 70–2
Widnall, Samuel Page *Mystery of Sixty Centuries, or a Modern St. George and the Dragon, A* 84
Wilberforce, Bishop Samuel 32, 36, 50
women, portrayal of 7, 37, 84–5, 90, 119, 126, 129, 131–2, 133–4, 135, 137, 138, 161, 178, 181, 185, 190, 191, 192, 201, *149, 159*
Wood, Lawson 121, 132–3, 181, 201

Zig Zag (revue) 162–4